MW00413088

Printed Circuit Board Design Techniques for EMC Compliance

Printed Circuit Board Design Techniques for EMC Compliance

Mark I. Montrose

Published under the
Sponsorship of the IEEE
Electromagnetic Compatibility Society

The Institute of Electrical and
Electronics Engineers, Inc.
New York
U.S.A.

IEEE PRESS
445 Hoes Lane, PO Box 1331
Piscataway, NJ 08855-1331

This book may be purchased at a discount from the publisher when ordered in bulk quantities. For more information, contact:

IEEE PRESS Marketing
Attn: Special Sales
P.O. Box 1331
445 Hoes Lane
Piscataway, NJ 08855-1331
Fax: (732) 981-9334

Printed in the United States of America

10 9 8 7 6 5 4

IEEE Order Number: PC5595

ISBN 0-7803-1131-0

Library of Congress Cataloging-in-Publication Data

Montrose, Mark I.
Printed circuit board design techniques for EMC compliance / Mark I. Montrose.
p. cm.
"Published under the sponsorship of the IEEE Electromagnetic Compatibility Society."
Includes bibliographical references and index.
ISBN 0-7803-1131-0 (cloth)
1. Printed circuits. 2. Electromagnetic compatibility. I. IEEE Electromagnetic Compatibility Society. II. Title.
TK7868.P7M66 1996 96-16931
621.3815'31—dc20 CIP

To
Margaret,
Maralena, and
Matthew

Contents

Preface

This design guide is presented to assist in printed circuit board design and layout, with the intent of meeting North American and international EMC compliance requirements. Many different layout design methodologies exist. This technical guide illustrates generally applicable layout methods for EMC compliance. Implementation of these methods may vary for a particular printed circuit board design.

The intended audience for this guide is engineers who design electronic products that use printed circuit boards. These engineers may focus on analog, digital, or system-level boards.

Regardless of their specialties, all engineers must produce a design that is suitable for actual production. Frequently, more emphasis is placed on functionality of the design than on overall system integration. System integration is usually assigned to product engineers, mechanical engineers, or others within the organization. Design engineers must now consider other aspects of product design, including the layout and production of printed circuit boards for EMC, which includes cognizance of the manner in which the electromagnetic fields transfer from the circuit boards to the chassis and/or case structure.

Not only must a design work properly, it must also comply with international regulatory requirements. Engineers who specialize in regulatory issues must evaluate products based on different standards. This guide presents techniques that will alleviate existing conflicts among various layout methods.

A great deal of technical information related to printed circuit board design and layout is available commercially as well as from public-domain documents. Typically, these sources provide only a brief discussion on how to implement a layout technique to solve an EMI problem (selected sources are listed in the Bibliography).

The guide itself is derived from lecture notes used to present EMC and printed circuit board layout information to engineers. The principles should prove useful for engineers in a variety of functions, including electrical and mechanical design, CAD/CAE, engineering and production test, manufacturing, and other fields.

Acknowledgments

Thanks to William (Bill) Kimmel and Daryl Gerke of Kimmel Gerke Associates, Ltd., St. Paul, Minnesota; Todd Hubing, University of Missouri; and Doug Smith, AT&T Bell Laboratories who provided technical review of the material for content and accuracy in addition to pointing out different aspects of printed circuit board design techniques not covered in my earlier drafts.

A special acknowledgment is given to Mr. W. Michael King of Costa Mesa, California, for his expertise, technical review, friendship, and encouragement without which I would never have achieved the technical knowledge to write this book.

My very special acknowledgment is to my wife, Margaret, and my two children, Maralena and Matthew, who tolerated my late night work and long hours at the keyboard. Without their understanding and support, this book could never have been written.

Mark I. Montrose
Santa Clara, California

1

Introduction

Printed Circuit Board Design Techniques for EMC Compliance is designed to help engineers minimize electromagnetic emissions generated by components (and circuits) to achieve acceptable levels of electromagnetic compatibility (EMC). It addresses both major aspects of EMC, which are

1. *Emissions:* Propagation of electromagnetic interference (EMI) from noncompliant devices (culprits), and in particular radiated and conducted radio frequency interference (RFI)
2. *Susceptibility:* The detrimental effects on susceptible devices (victims) of EMI in forms that include electrostatic discharge (ESD) and other forms of electrical overstress (EOS)

The engineer's goal is to meet design requirements to satisfy both international and domestic regulations and voluntary industry standards.

The information presented in this volume is focused on "non-EMC" engineers who design and lay out printed circuit boards (PCBs). EMC engineers will also find the information helpful in solving design problems at the PCB level. This guideline is applicable as a reference document throughout any design project.

Circuit technology is advancing rapidly, and design techniques that worked several years ago are no longer effective in today's high-speed digital products. Because EMC is insufficiently covered in engineering schools, training courses and seminars are held throughout the country to provide this information. As such, there is a widespread need for introductory material. With this in mind, *Printed Circuit Board Design Tech-*

niques for EMC Compliance is written for engineers who never studied applied electromagnetics in school or who have limited hands-on experience with high-speed, high-technology printed circuit board design as it specifically relates to EMC compliance.

A minimal amount of mathematical analysis is presented herein. It is the intent of this guideline to describe *hands-on techniques* that have been successfully applied to many real-world products. Data is presented in a format that is easy to understand and implement. Those interested in Maxwell's equations or the more highly technical aspects of PCB design theory will find a list of appropriate materials in the bibliography.

The focus of this guideline is *strictly* on the printed circuit board. Discussion of containment techniques (box shielding), internal and external cabling, power supply design, and other system-level subassemblies that use printed circuit boards as subcomponents will not be thoroughly discussed. Again, excellent reference material is listed in the bibliography on these aspects of EMC system design engineering.

Controlling emissions has become a necessity for acceptable performance of an electronic device in both the civilian and military environment. It is more cost-effective to design a product with suppression on the printed circuit board than to "build a better box." Containment measures are not always economically justified and may degrade as the EMC life cycle of the product is extended beyond the original design specification. For example, end users often remove covers from enclosures for ease of access during repair or upgrade. Sheet metal covers (particularly internal subassembly covers that act as partition shields) in many cases are never replaced. The same is true for blank metal panels or faceplates on the front of a system that contains a chassis or backplane assembly. As a result, containment measures are compromised. Proper layout of a printed circuit board with suppression techniques also assists in EMC compliance at the level of cables and interconnects, whereas box shielding (containment) does not.

While it is impossible to anticipate every application or design concern possible, this book provides details on how to implement a variety of design techniques for most applications. The concepts presented are *fundamental* in nature and are applicable to all electronic products. While every design is different, the basic fundamentals of product design rarely change, and EMC theory is constant.

Why worry about EMC compliance? After all, isn't speed the most important design parameter? Legal requirements dictate the maximum permissible interference potential of digital products. These requirements are based on experience in the marketplace related to emission and immunity complaints. Often, these same techniques will aid in improving signal quality and signal-to-noise performance.

This text discusses high-technology, high-speed designs that require new and expanded techniques for EMC suppression at the PCB level. Many techniques that were used successfully several years ago are now less effective for proper signal functionality and compliance. Components have become faster and more complex. Use of custom gate array logic and ASICs presents new and challenging opportunities for EMC engineers. The design and layout of a printed circuit board for EMI suppression at the source must always be optimized while maintaining system-wide functionality.

1.1 FUNDAMENTAL DEFINITIONS

The following basic terms are used throughout this book:

Electromagnetic compatibility (EMC)—The ability of a product to coexist in its intended electromagnetic environment without causing or suffering functional degradation or damage.

Electromagnetic interference (EMI)—A process by which disruptive electromagnetic energy is transmitted from one electronic device to another via radiated or conducted paths (or both). In common usage, the term refers particularly to RF signals, but EMI can occur in the frequency range "from dc to daylight."

Radio frequency (RF)—The frequency range within which coherent electromagnetic radiation is useful for communication purposes—roughly from 10 kHz to 100 GHz. This energy may be generated intentionally, as by a radio transmitter, or unintentionally as a by-product of an electronic device's operation. RF energy is transmitted through two basic modes:

- *Radiated emissions (RE)*—The component of RF energy that is transmitted through a medium as an electromagnetic field. RF energy is usually transmitted through free space; however, other modes of field transmission may occur.
- *Conducted emissions (CE)*—The component of RF energy that is transmitted through a conductive medium as an electromagnetic field, generally through a wire or interconnect cables. Line conducted interference (LCI) refers to RF energy in a power cord.

Susceptibility—A relative measure of a device or system's propensity to be disrupted or damaged by EMI exposure.

Immunity—A relative measure of a device or system's ability to withstand EMI exposure.

Electrical overstress (EOS)—Damage or loss of functionality experienced by an electronic device as a result of a high-voltage pulse. EOS includes lightning and electrostatic discharge events.

Electrostatic discharge (ESD)—A high-voltage pulse that may cause damage or loss of functionality to susceptible devices. Although lightning qualifies as a high-voltage pulse, the term ESD is generally applied to events of lesser amperage, and more specifically to events that are triggered by human beings. However, for the purposes of this text, lightning will be included in the overall ESD category because the protection techniques are very similar, although differing in magnitude.

Radiated susceptibility—The relative inability of a product to withstand EMI that arrives via free-space propagation.

Conducted susceptibility—The relative inability of a product to withstand electromagnetic energy that reaches it through external cables, power cords, and other I/O interconnects.

Containment—Preventing RF energy from exiting an enclosure, generally by shielding a product within a metal enclosure (Faraday cage) or by using a plastic housing with RF conductive paint. By reciprocity, we can also speak of containment as preventing RF energy from entering the enclosure.

Suppression—Designing a product to reduce or eliminate RF energy at the source without relying on a secondary method such as a metal housing or chassis.

The next section discusses the nature of EMC, its relation to printed circuit boards, and the legal mandates for EMC compliance.

1.2 EMC AND THE PRINTED CIRCUIT BOARD

Traditionally, EMC has been considered an art of "black magic." In reality, EMC can be explained by mathematical concepts. Some of the relevant equations and formulas are complex and beyond the scope of this design guideline. Fortunately, simple models can be formulated to describe how and why EMC can be achieved.

Many variables exist in the creation of EMI. This is because EMI is often the result of exceptions to the normal rules of passive component behavior. A resistor at high frequencies acts like a series combination of inductance with resistance in parallel with a capacitor. A capacitor at high frequencies acts like an inductor and resistor in a series combination with the capacitor. An inductor at high frequencies performs like an inductor and capacitor in parallel. An illustration [1] of these abnormal behaviors of passive components at both high and low frequencies is shown in Fig. 1.1.

These behavioral characteristics are referred to as the "hidden schematic." Digital engineers generally assume that components have a single frequency response. As a result, passive component selection is based on functional performance in the time domain without regard to the real characteristics exhibited in the frequency domain. Many times, EMI exceptions occur if the designer bends or breaks the rules, as seen in Fig. 1.1.

To restate the complex problems that exist, consider the field of EMC as "*everything that is not on a schematic or assembly drawing*." This statement is why the field of EMC is sometimes considered to be an art of black magic.

Once the hidden behavior of components is understood, it becomes a simple process to design products with circuit boards that pass EMC requirements. Hidden behavior also takes into consideration the switching speed of active components along with their unique characteristics, which also may have hidden resistive, capacitive, and inductive components.

Designing products that will pass legally required EMI tests is not as difficult as one might expect. Engineers strive to design elegant products, but elegance sometimes must give way to other engineering considerations such as product safety, manufacturing cost, and, of course, EMC. Such abstract problems can be challenging, particularly if the engineer is unfamiliar with the types and levels of compliance required. The general

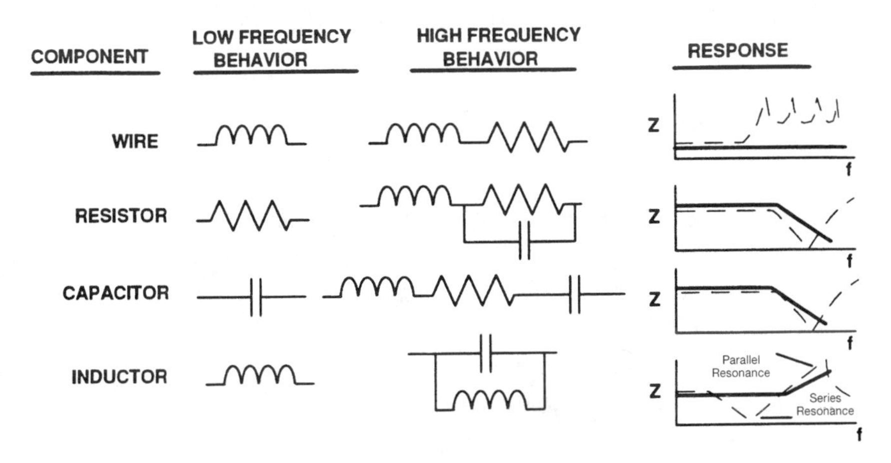

Fig. 1.1 Component Characteristics at RF Frequencies (Source: Designers Guide to Electromagnetic Compatibility, *EDN.* © 1994, Cahners Publishing Co. Reprinted by permission.)

guidelines offered in this book will remove the mystery from the "*hidden schematic*."

When an EMI problem occurs, the engineer should approach the situation logically. A simple model that describes the field of EMC has three elements:

1. a source of energy
2. a receptor that is disrupted by this energy
3. a coupling path between the source and receptor

For interference to exist, all three elements must be present. If one of the three elements is removed, there is no interference. It therefore becomes our first task to determine which is the easiest element to remove. Generally, designing a printed circuit board that eliminates most sources of RF interference is the most cost-effective approach (called *suppression*). The second and third elements tend to be addressed with containment techniques. Figure 1.2 illustrates the relationship between these three areas and presents a list of products typically associated with each element.

A product must be designed for two levels of performance: one to minimize RF energy exiting an enclosure (emissions), and the other to minimize the amount of RF energy entering the enclosure (susceptibility or immunity). In both cases, there are considerations for radiated and conducted EMI. This relationship is shown in Fig. 1.3.

When dealing with emissions, the general rule of thumb is:

> *The higher the frequency, the greater the likelihood of a radiated coupling path; the lower the frequency, the greater the likelihood of a conducted coupling path.*

There are five major considerations in EMI analysis, as enumerated below:

1. *Frequency.* Where in the frequency spectrum is the problem observed?
2. *Amplitude.* How strong is the source energy level, and how great is its potential to cause harmful interference?
3. *Time.* Is the problem continuous (clock signals), or does it exist only during certain cycles of operation (i.e., disk drive write operation)?
4. *Impedance.* What is the impedance of both the source and receptor units and the impedance of the transfer mechanism between the two?

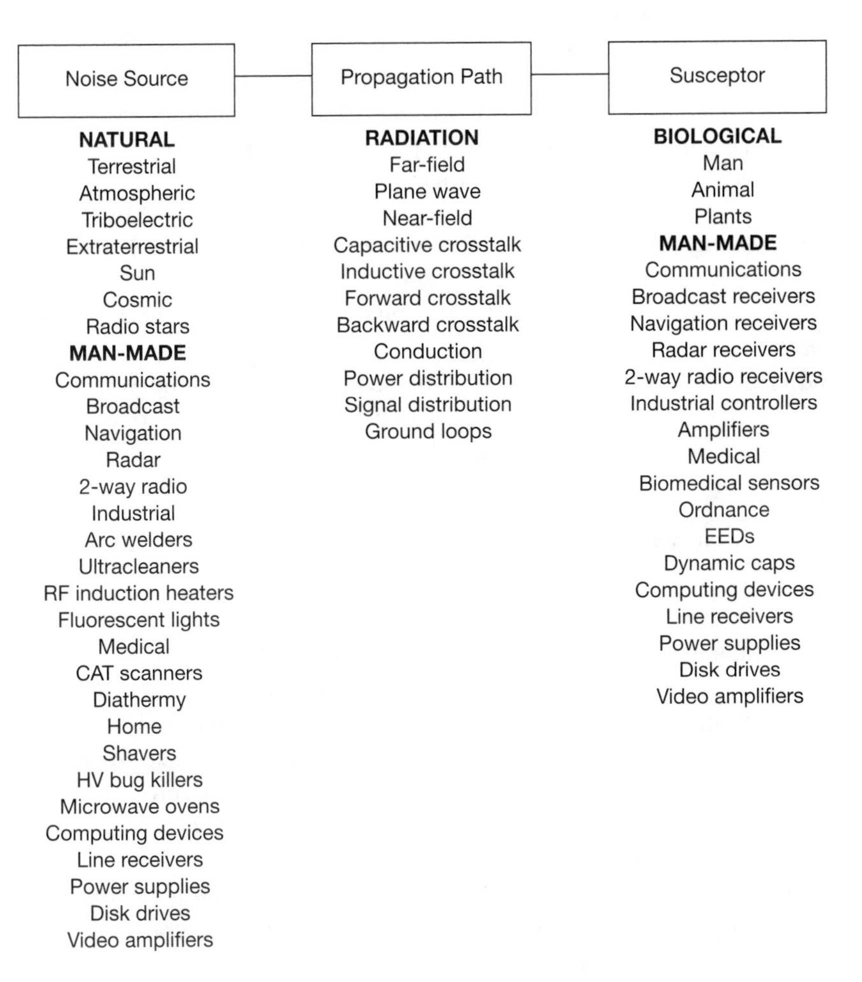

Fig. 1.2 Elements of an EMC Environment

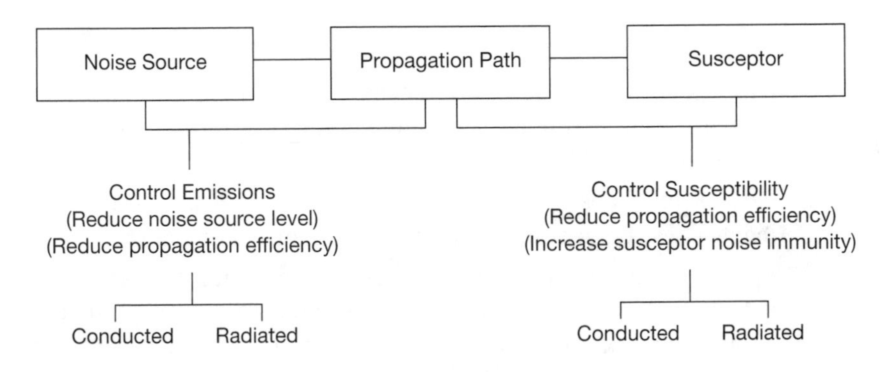

Fig. 1.3 Coupling Paths

5. *Dimensions.* What are the physical dimensions of the emitting device? RF currents will exit an enclosure through chassis leaks that equal fractions of a wavelength or significant fractions of a "rise-time distance." Trace lengths on a printed circuit board are also transmission paths for RF currents.

Regarding impedance, if both source and receptor have the same impedance, one should expect greater emission problems than if the source and receptor have different impedances. This is because high-impedance sources have minimal impact on low-impedance receptors, and vice-versa. Similar rules apply to radiated coupling. High impedances are associated with electric fields, whereas low impedances are associated with magnetic fields.

1.3 NORTH AMERICAN REGULATORY REQUIREMENTS

Electrical and electronic products generate RF energy. Emission levels are set by rules and regulations as mandated by national and international organizations. In the U.S.A., the Federal Communications Commission (FCC) regulates radio and wire communications. In Canada, the Department of Communication (DOC) performs the same function. Internationally, each country has a designated agency within its government to oversee all aspects of communication.

The FCC regulates electronic products by specifying technical standards and operational requirements in the Code of the Federal Regulations (CFR), Title 47. The sections applicable to products discussed herein are Parts 15, 18, and 68. These regulations have been developed over many years and are based on complaints filed with the Commission. The most prominent are listed below. In Canada, the specification equivalent to CFR 47, Part 15 is SOR 88/475.

1. Part 15 is applicable to unlicensed radio-frequency radiating devices (both intentional and unintentional). Information technology equipment (ITE) falls within Part 15.
2. Part 18 regulates industrial, scientific, and medical (ISM) equipment. These devices use radio waves for normal operation.
3. Part 68 regulates electronic equipment connected to a telephone network. This Part provides a uniform standard to protect the telephone network from harm caused by terminal equipment connected to it.

The FCC and DOC define a digital device as:

> An unintentional radiator (device or system) that generates and uses timing signal pulses at a rate in excess of 9,000 pulses (cycles) per second and uses digital techniques; inclusive of telephone equipment that uses digital techniques or any device or system that generates and utilizes radio frequency energy for the purpose of performing data processing functions, such as electronic computations, operations, transformation, recording, filing, sorting, storage, retrieval or transfer.

Digital computing products are classified into two Categories: Class A and B. The FCC and DOC use the same definitions:

Class A: A computing device that is marketed for use in a commercial, industrial, or business environment, exclusive of a device which is marketed for use by the general public or is intended to be used in the home.

Class B: A computing device that is marketed for use in a residential environment, notwithstanding its use in a commercial, industrial or business environment.

If a product contains digital circuitry and has a clock frequency greater than 9 kHz, it is defined as a *digital device* and is subject to rules and regulations of the FCC and DOC. Electromagnetic interference may occur due to both time-domain and frequency-domain components of both digital and analog circuits. These products are subject to both domestic and international regulations.

The FCC and DOC regulate conducted emissions on power cords (line conducted interference) from 450 kHz to 30 MHz. Radiated emissions are measured from 30 MHz to 1000 MHz.

1.4 WORLDWIDE REGULATORY REQUIREMENTS

Harmonization of test requirements, standards, and procedures is being implemented on a worldwide basis. Principles discussed herein will allow regulatory compliance to be achieved with minimal development costs and shorter design cycles. The harmonization process is based on the work of expert technical committees reporting to the International Electrotechnical Commission (IEC). The Committee for European Electrotechnical Standardization (Comité Européen de Normalisation Electrotechnique, or CENELEC) adopts standards developed by the IEC and the International Special Committee on Radio Interference (Comité International Spécial des Perturbations Radioélectriques, or CISPR). CISPR is not a government body, but a private organization. CISPR is a subcommittee of the IEC. CISPR issues recommendations to the IEC for adoption. The IEC presents these recommendations to CENELEC which, in turn, sends them on to the European Parliament for adoption. Once adopted by the European Parliament, it is the responsibility of each member country of the European Union (EU) to adopt these requirements into their national law.

It is common to refer to international specifications as CISPR when, in fact, the real standards, after adoption and publication by the European Parliament, are prefixed with an EN number (European Normalisation). To summarize, the European Parliament places into law requirements developed by CISPR and other European working groups and committees under the auspices of CENELEC.

This book "focuses on products that fall within the category of information technology equipment" and are covered by EN 55 022. CISPR regulates conducted emissions on power cords from 150 kHz to 30 MHz. Radiated emissions are measured from 30 MHz to 1000 MHz.

The most commonly referenced CISPR test standards for products that contain printed circuit boards are listed below. This book is applicable to these commonly referenced standards. Many other test standards exist. This list is subject to periodic change due to continuing development in standards writing, along with continuing harmonization within the European Union (EU). The EU was formerly known as the European Community (EC) or European Economic Community (EEC). This list is current at date of publication and are subject to change without notice.

EN 50 081-1: 1992
: Electromagnetic compatibility generic emission standard—Part 1: Residential, commercial and light industry

EN 50 081-2: 1994
: Electromagnetic compatibility generic emission standard—Part 2: Heavy industrial environment

EN 50 082-1: 1993
: Electromagnetic compatibility generic immunity standard—Part 1: Residential, commercial and light industry

EN 50 082-2: 1994
: Electromagnetic compatibility generic immunity standard—Part 2: Heavy industrial environment

EN 55 011: 1991
Limits and methods of measurements of radio disturbance characteristics of industrial, scientific and medical (ISM) radio-frequency equipment (CISPR 11: 1990 ed. 2)

EN 55 013: 1993
Limits and methods of measurements of radio disturbance characteristics of broadcast receivers and associated equipment (CISPR 13: 1975 ed. 1 + Amendment 1: 1992)

EN 55 014: 1993
Limits and methods of measurements of radio disturbance characteristics of household electrical appliances, portable tools and similar electrical apparatus (CISPR 14: 1993 ed. 3)

EN 55 020: 1993
Limits and methods of measurements of radio disturbance characteristics of broadcast receivers and associated equipment (CISPR 20: 1990 ed. 2 + Amendment 1: 1990)

EN 55 022: 1994
Limits and methods of measurements of radio disturbance characteristics of information technology equipment (CISPR 22: 1985 ed. 1)

Products are classified into two categories of emissions: Class A and Class B. CISPR defines these categories as:

Class A: Equipment is information technology equipment if it satisfies the Class A interference limits but does not satisfy the Class B limits. In some countries, such equipment may be subjected to restrictions on its sale and/or use.
(Note: The limits for Class A equipment are derived for typical commercial establishments for which a 30 m protection distance is used. The class A limits may be too liberal for domestic establishments and some residential areas).

Class B: Equipment is information technology equipment if it satisfies the Class B interference limits. Such equipment should not be subjected to restrictions on its sale and is generally not subject to restrictions on its use.
(Note: The limits for Class B equipment are derived for typical domestic establishments for which a 10 m protection distance is used).

Limits for European standards are similar but different from the North American requirements. Appendix B illustrates the specification limits for

FCC/DOC (Part 15/SOR 88/475) and various international standards in both tabular and graphical format.

International standards for susceptibility (immunity) are provided in the EN 61000-4-X series. This series describes the test and measurement methods of basic standards. Basic standards are specific to a particular type of EMI phenomenon. It is not limited to a specific type of product. Internal to this series are the following:

- terminology
- descriptions of the EMI phenomenon
- instrumentation
- measurement and test methods
- ranges of severity levels with regard to the immunity of the equipment

The EN 61000-4-X series is based on the well known IEC 801-X and IEC 1000-4-X requirements. The main difference is in the title and publication number. Future changes in technical requirements may be significant between the EN 61000-4-X and the IEC 801-X series. IEC requirements were officially withdrawn from circulation and replaced with a newly designated series. The subparts of IEC 1000-4-X are listed below. As of the date of writing, only IEC 1000-4-2/3/4 are legally required for EMC compliance. It is anticipated that in the future, additional test standards and requirements will be mandatory. Currently, immunity tests are mandated in Europe, only recommended in North America, and optional worldwide.

- IEC 1000-1 General Considerations
- IEC 1000-2 Environment
- IEC 1000-3 Limits/Generic Standards
- IEC 1000-4 Test and Measurement Techniques
    - EN 61000-4-1 Administrative Aspects of the Directive
    - EN 61000-4-2 Electrostatic Discharge
    - EN 61000-4-3 Radiated Susceptibility
    - EN 61000-4-4 Electrical Fast Transients
    - EN 61000-4-5 Surge Voltage Immunity*
    - EN 61000-4-6 Conducted Disturbance Induced by RF Fields Above 9 kHz

* Indicates that the standard, *as of the date of writing*, is in draft form, published but not official, or not yet legally required as part of the EMC directive. The reader should verify the existence and status of a particular IEC 1000-4-X document before implementing tests.

EN 61000-4-7 Harmonics, Interharmonics and Instrumentation for Power Supply Systems*

EN 61000-4-8 Power Frequency Magnetic Field Immunity

EN 61000-4-9 Pulsed-Magnetic Field Immunity

EN 61000-4-10 Damped Oscillatory Magnetic Field

EN 61000-4-11 Voltage Dips, Short Interrupts and Voltage Variations Immunity

EN 61000-4-12 Oscillatory Waves*

- IEC 1000-5 Installations and Mitigation Guideline
- IEC 1000-6 Miscellaneous

To summarize, the standards that are required now and in the future for compliance with the EMC Directive 89/336/EEC, amended in 1992, are contained in Table 1.1, even though some of these test standards have not yet been adopted and published. Until the test requirements are officially published in the European Official Journal, compliance is optional.

1.5 ADDITIONAL NORTH AMERICAN REGULATORY REQUIREMENTS

Other agency requirements in North America include those listed in Table 1.2. These standards are very specific and for the most part beyond the scope of this design guideline. The list is presented for completeness only, given that printed circuit boards are used in products covered by these standards.

Table 1.1 International Emissions and Immunity Standards

Emissions	*Immunity*
EN 55 022 (ITE) Information technology equipment	IEC 1000-4-2 ESD IEC 1000-4-3 Radiated susceptibility IEC 1000-4-4 Electrical fast transients
EN 55 014 (HHA) Household appliances, hand tools, and similar apparatus	IEC 1000-4-5 Lightning on power lines IEC 1000-4-6 Conducted continuous wave IEC 1000-4-8 Magnetic radiated fields
EN 55 011 (ISM) Industrial, scientific, and medical equipment	IEC 1000-4-11 Sags, surges, dropouts
EN 60 555-2 Power line harmonics EN 60 555-3 Power line flicker	

Table 1.2 Additional North American Standards

Standard	*Subject Area*
MDS-201-0004 FDA Standard for Medical Devices	Emissions and susceptibility in medical electronics
SAE J 551	Radiated EMI from vehicles and associated devices
NACSIM 5100 (a.k.a. TEMPEST)	Classified standard limiting emissions from certain products to be sufficiently low to prevent interception and deciphering data streams that contain intelligence
MIL-STD-461/462	U.S. military emission standards and test procedures, both radiated and conducted

1.6 SUPPLEMENTAL INFORMATION

In addition to compliance for EMC, requirements exist for product safety. These requirements include energy hazards and flammability. Many printed circuit boards are subject to high voltage and current levels that pose a possible shock hazard to the user. In addition, excessive current flow on traces generates heat, which can cause the fiberglass material used in the construction of the printed circuit board to burn and/or melt, with an associated risk of fire. Components and interconnects on a printed circuit board may also provide a source of fuel (combustible material) that may contribute to a fire hazard under abnormal fault conditions.

An important part of this design guide is the Appendix. Much technical information is contained in all the chapters. To assist during the actual layout of a printed circuit board, Appendix A, Summary of Design Techniques, provides a brief overview of items discussed, cross-referenced to their respective chapters. This summary may be used for quick review during the layout and design stage.

Appendix B is provided as a quick reference guide to international EMC specification limits for the United States and Canada (FCC/DOC), Europe, and worldwide, in addition to the European immunity limits.

1.7 REFERENCE

1. Gerke, D., and W. Kimmel. 1994. The Designers Guide to Electromagnetic Compatibility. *EDN* (January 20).

2

Printed Circuit Board Basics

When designing a printed circuit board (PCB), one of the first considerations is how many routing layers and power planes are required for functionality (within the context of acceptable costs). The number of layers is determined by functional specification, noise immunity (use of power planes) and signal category separations, number of nets (traces) to be routed, impedance control, component density of VLSI circuits, routing of buses, and the like. Proper use of stripline and microstrip topology are required for radio frequency (RF) suppression in the PCB. It is desirable to suppress RF energy on the PCB rather than to rely on containment by a metal chassis or conductive plastic enclosure. The use of planes (voltage and ground) embedded in the PCB is one of the most important methods of suppressing common-mode RF internal to the board. The advantage over most other design techniques is that these planes intrinsically contribute to reducing high-frequency power distribution impedance.

Figure 2.1 illustrates the difference between microstrip and stripline topologies, which represent the two major classifications. Each is described below:

- Microstrip refers to outer trace(s) on a PCB, which are separated by a dielectric material and then a solid plane. Although microstrip techniques provide suppression of RF energy on the board, faster clock and logic signals are possible than with stripline. The faster signals are due to less capacitive coupling and lower unloaded propagation delay between traces routed on the outer layers and an adjacent plane. Capacitors are sometimes used on clock signals to smooth the edges of fast signals. With lower capacitive coupling

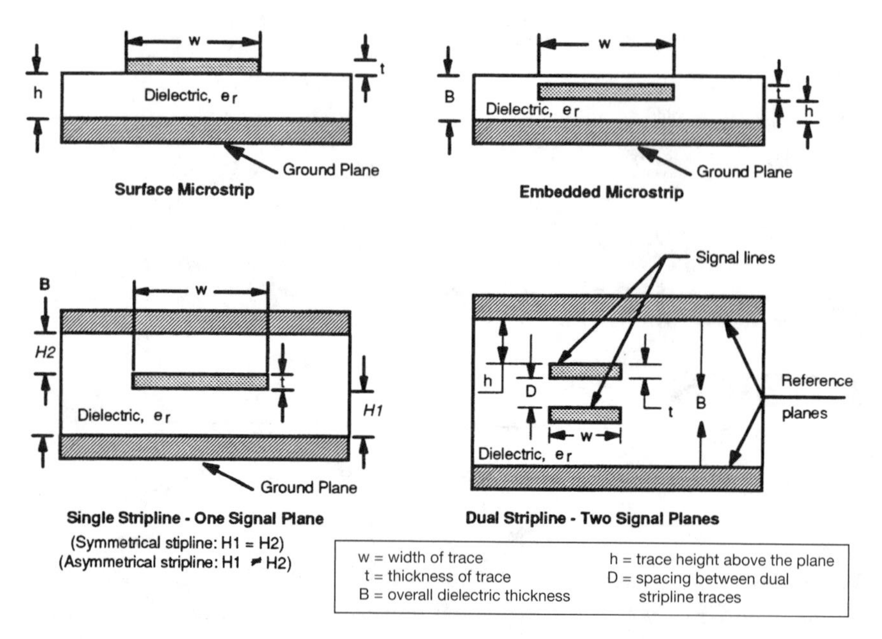

Fig. 2.1 Microstrip and stripline topologies

between two solid planes, faster signal propagation can be implemented. The drawback of microstrip is that the outer layers of the PCB can radiate RF energy into the environment unless one adds the protection of a plane on both sides of this outer circuit plane (i.e., shielding of both the top and bottom sides of the traces).

- Stripline refers to placement of a circuit plane between two solid planes—either voltage or ground. Stripline provides better noise immunity for RF emissions, but it comes at the expense of slower propagation speeds. Because the circuit (signal) plane is located between solid planes (power or ground), capacitive coupling will occur between these two planes, which slows down the edge rates of high-speed signals. Capacitive coupling effects on stripline topology are generally observed on signals with edges faster than 1 ns. The main benefit of using stripline is the complete shielding of RF energy generated from internal traces and the consequent suppression of RF radiation. This phenomenon is discussed further in various chapters throughout this guide.

One item to note is that radiation may still occur from certain other components. Although the internal signal traces may not radiate RF energy, the interconnects (bond wires, lead frames, sockets, interconnect cables, and the like) still pose problems. Depending on the impedance of the system, components, and traces, an impedance mismatch may exist. This mismatch may couple RF energy from internal circuit traces to other circuits or free space. Minimizing lead inductance from components on the outer layers of the PCB will reduce radiated emission effects.

2.1 LAYER STACKUP ASSIGNMENT

The following assignments are provided as a *guide* to the selection of a stackup method for PCBs. These assignments are *not* cast in stone and may be modified as appropriate for functionality and number of routing layers required. The important concept to observe is that *each and every routing layer must be adjacent to a solid plane (power or ground).* A summary of these assignments is presented in Table 2.1.

Table 2.1 Example of Stackup Assignment

Layer #	*1*	*2*	*3*	*4*	*5*	*6*	*7*	*8*	*9*	*10*	*Comments*
2 layers	S1 G	S2 P									Lower-speed designs
4 layers (2 routing)	S1	G	P	S2							Difficult to maintain high signal impedance *and* low power impedance
6 layers (4 routing)	S1	G	S2	S3	P	S4					Lower-speed design, poor power high signal impedance
6 layers (4 routing)	S1	S2	G	P	S3	S4					Default critical signals to S2 only
6 layers (3 routing)	S1	G	S2	P	G	S3					Default lower-speed signals to S2–S3
8 layers (6 routing)	S1	S2	G	S3	S4	P	S5	S6			Default high-speed signals to S2–S3. It has poor power impedance
8 layers (4 routing)	S1	G	S2	G	P	S3	G	S4			Best for EMC
10 layers (6 routing)	S1	G	S2	S3	G	P	S4	S5	G	S6	Best for EMC. S4 is susceptible to power noise

S = signal routing layer, P = power, G = ground

2.1.1 Two-layer boards

There are two layout methodologies for two-layer boards. The first is used for older technologies (low-speed components) that generally consist of dual in-line (DIP) packages in a straight row or matrix configuration. Very few engineers use this technique today. Configuration 1 is presented for completeness only. The second configuration is typical of current design practice.

Configuration 1 (Fig. 2.2)

- Layer the power and ground in a grid style with the total loop area formed by each grid square not exceeding 1.5 square inches
- Run power and circuit traces at a 90° angle to each other, with power on one layer, ground on the other layer.
- Place ground traces on the top layer, vertical polarization. Place power traces on the bottom layer, horizontal polarization.
- Locate decoupling capacitors between the power and ground traces at all connectors and at each IC.

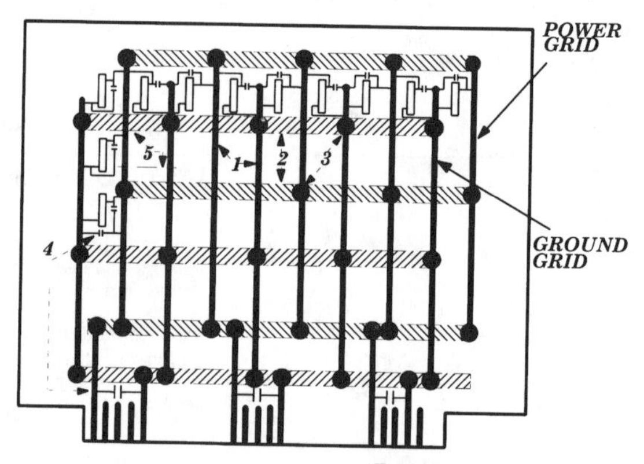

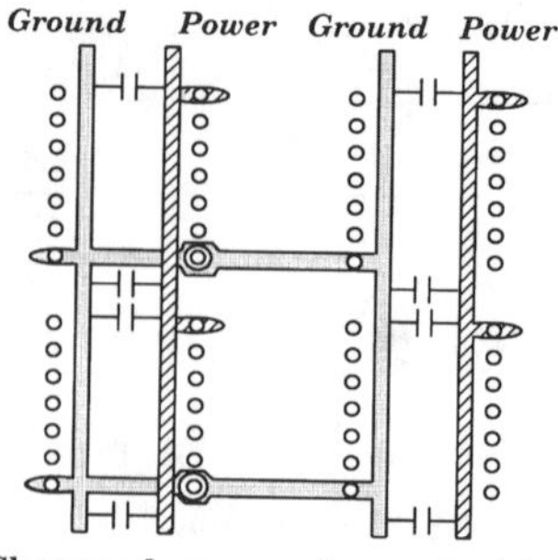

Fig. 2.2 Two-layer PCB with power grid

Configuration 2 (Fig. 2.3)

This configuration is commonly used in low-frequency analog designs running at less than 10 kHz.

- Route all power traces in a *radial fashion* from the power supply to all components on the same routing layer. Minimize the total length of all traces.
- Route all ground and power traces adjacent to (in parallel with) each other. This minimizes loop currents that may be created by high-frequency switching noise (internal to the components) from corrupting other circuits and control signals. The only time these traces are to be separated by a distance less than the width of any individual trace is when they must separate for connection to the decoupling capacitor. *Signal flow should parallel these ground paths.*
- Avoid tying different branches of a "tree" to other branches, thereby preventing ground loops from being created.

In examining Fig. 2.3, observe that low-frequency parasitic inductance and capacitance generally do not cause problems. For this situation, sin-

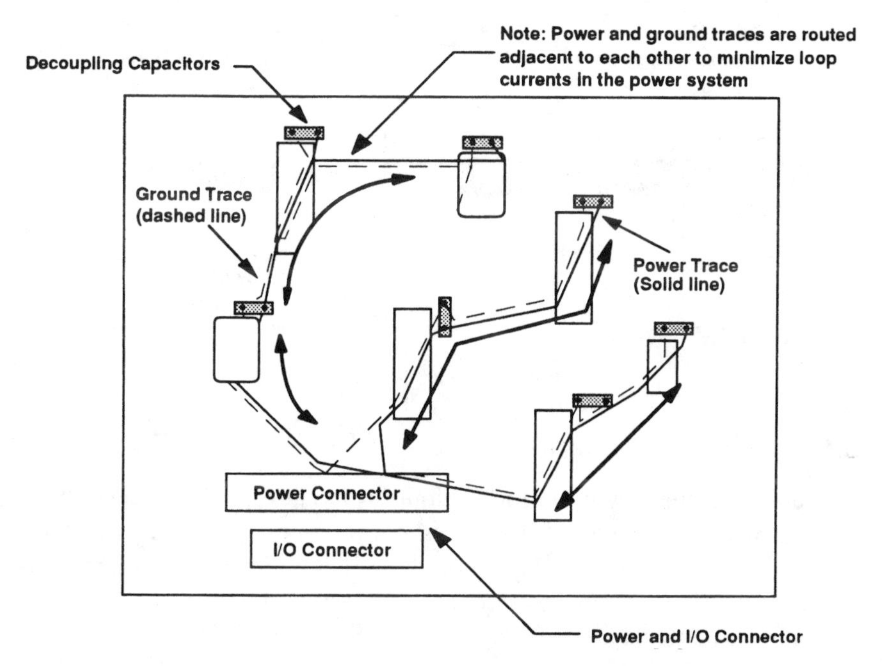

Fig. 2.3 Two-layer PCB with radial structure for power routing and signal flow migration

gle-point grounding is recommended. In other words, implement high-frequency performance in layout for a low-frequency application.

In Fig. 2.3, note the following:

- For high-frequency applications, control the surface impedance (Z) of all signal traces and their return current paths.
- When used in a low-frequency application, control topology layout rather than impedance.

The reason components are placed on a two-layer board as shown in Fig. 2.3 is illustrated in Fig. 2.4. This deals with radial migration from high-bandwidth (CPU) devices to lower-speed components (I/O). What *radial migration*[*] means is that, as circuits progress from high-bandwidth to low-bandwidth areas, a slowing down of the signal propagation delay of the traces occurs, with enhanced EMI performance at the I/O connector. This signal propagation slowdown occurs because devices have internal capacitance and propagation delay. Each device slows down the edge rate, t_r, of the signal. As circuits cascade from the CPU section to I/O, this delay is summed up (like a filter) until the high-bandwidth components are removed from the system and I/O circuits.

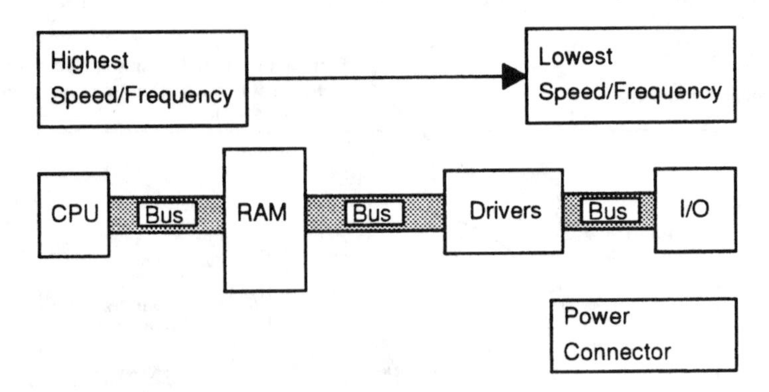

Fig. 2.4 Radial migration

2.1.2 Four-layer boards

There is only one way to perform a four-layer stackup. Use of power and ground planes enhances EMI suppression in comparison to that of two-layer boards; however, four-layer boards are not optimal for flux cancellation of RF currents created by circuits and traces. Figure 2.5 illustrates this stackup in greater detail.

[*] Technique developed by W. Michael King

- first layer (component side), signals and clocks
- second layer, ground plane
- third layer, power plane
- fourth layer (solder side), signals and clocks

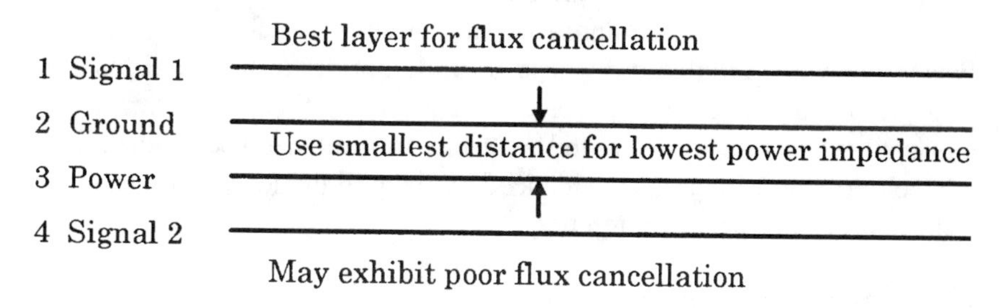

Fig. 2.5 Four-layer board stackup

For the following stackup assignments, it is observed that where three or more planes are provided (i.e., one power and two ground planes), optimal performance of extra-high-speed clock traces are achieved when they are routed "adjacent" to a ground plane and "not adjacent" to the power plane. This is one of the *basic fundamental concepts* of EMI suppression in a PCB, and it should become fixed firmly in the reader's mind.

> Multilayer boards provide superior signal quality EMC performance because signal impedance control through stripline or microstrip is observed. The distribution impedance of the power and ground planes must be dramatically reduced. These planes contain RF spectral current surges caused by "logic crossover," momentary shorts, and capacitive loading on signals with wide buses. Central to the issue of microstrip (or stripline) application is understanding flux cancellation that minimizes (controls) inductance in any transmission line. Various logic devices may be quite asymmetrical in their pull-up/pull-down current ratios. This means that flux cancellation is enhanced between the signal and the ground planes rather than the power planes. With this situation, use of the power plane as a flux cancellation control may not present an optimum condition, resulting in signal flux phase shift, greater inductance, poor impedance control, and noise instability. Use of the ground plane for optimal signal reference is thus preferred [1].

To briefly restate this important concept of PCB flux cancellation, it is noted that not all components behave the same on the board in relation to their pull-up/pull-down current ratios. For example, some devices have 15 mA pull-up/65 mA pull-down. Other devices have 65 mA pull-up/pull-down. An asymmetrical condition exists that creates an imbalance in the power and ground plane structure. The fundamental concept of board-level suppression lies in flux cancellation between RF currents that exist within the board traces, components, and circuits, in relation to a plane. Power planes, due to this flux phase shift, do not perform as well for flux cancellation as do ground planes. As a result, optimal performance is achieved when traces are routed adjacent to ground planes rather than adjacent to power planes, as evidenced by the pull up/down ratios that are indicative of flux-phase preference.

2.1.3 Six-layer boards

Three common configurations are used for six-layer PCBs.

Configuration 1. This is commonly used with clock signals or high frequency components (see Fig. 2.6).

- first layer (component side), microstrip signal routing layer
- second layer, ground plane
- third layer, stripline routing layer
- fourth layer, stripline routing layer
- fifth layer, power plane
- sixth layer (solder side), microstrip signal routing layer

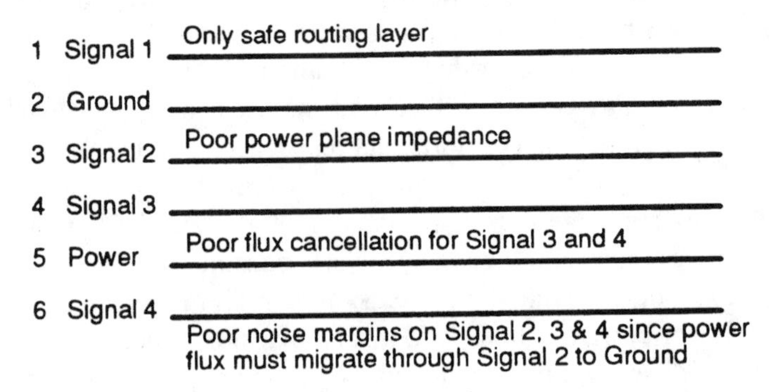

Fig. 2.6 Six-layer PCB, configuration 1

Configuration 2. This arrangement offers improved performance due to increased planar decoupling between voltage and ground—four routing layers (see Fig. 2.7).

- first layer (component side), microstrip signal routing layer
- second layer, embedded microstrip routing layer
- third layer, ground plane
- fourth layer, power plane
- fifth layer, embedded microstrip routing layer
- sixth layer (solder side), microstrip signal routing layer

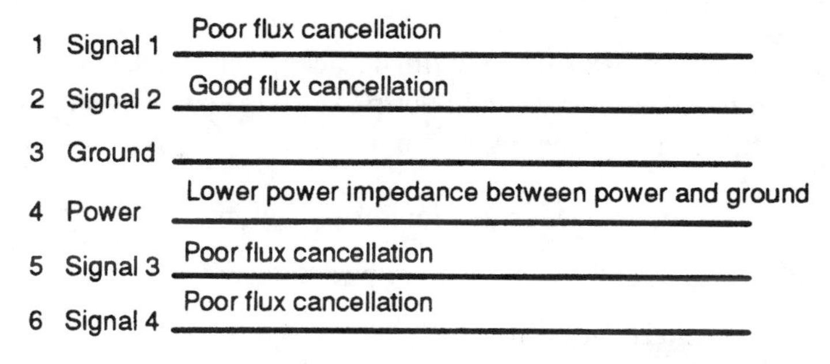

Fig. 2.7 Six-layer PCB, configuration 2

Configuration 3. This offers the best performance with increased flux cancellation for all routing layers and lower power plane impedance—three routing layers (see Fig. 2.8).

- first layer (component side), microstrip signal routing layer
- second layer, ground plane
- third layer, stripline routing plane, followed by fill material
- fourth layer, power plane
- fifth layer, ground plane
- sixth layer (solder side), microstrip signal routing layer

1 Signal 1 — Excellent routing layer (X)
2 Ground — Good flux cancellation — X-Y paired traces
3 Signal 2 — Excellent routing layer (Y)
Fill material
4 Power
5 Ground — Lower power impedance
6 Signal 3 — Good flux cancellation

Fig. 2.8 Six-layer PCB, configuration 3

2.1.4 Eight-layer boards

Two types of assignments generally are employed. The first configuration provides minimal EMI flux cancellation. The second configuration provides maximum cancellation due to use of additional solid planes and tighter flux cancellation for RF currents. Determination of whether to use configuration 1 or 2 depends on the number of nets to be routed, component density (pin count), size of bus structures, analog and digital circuitry, and available real estate.

Configuration 1. This is not an optimal stackup scheme due to poor flux cancellation on signal planes and poor power impedance. It employs six routing layers and two planes as shown in Fig. 2.9.

- first layer (component side), microstrip routing signal layer
- second layer, embedded microstrip routing signal layer
- third layer, ground plane
- fourth layer, stripline routing signal layer
- fifth layer, stripline routing signal layer
- sixth layer, power plane
- seventh layer, embedded microstrip routing signal layer
- eighth layer (solder side), microstrip routing signal layer

1 Signal 1
2 Signal 2 — Excellent routing layer (X)
3 Ground — X-Y paired traces
4 Signal 3 — Excellent routing layer (Y)
5 Signal 4 — Poor flux cancellation
6 Power
7 Signal 5 — Poor noise margin on Signal 5 & 6 since power flux must migrate
8 Signal 6 — through Signal 3 & 4 to ground

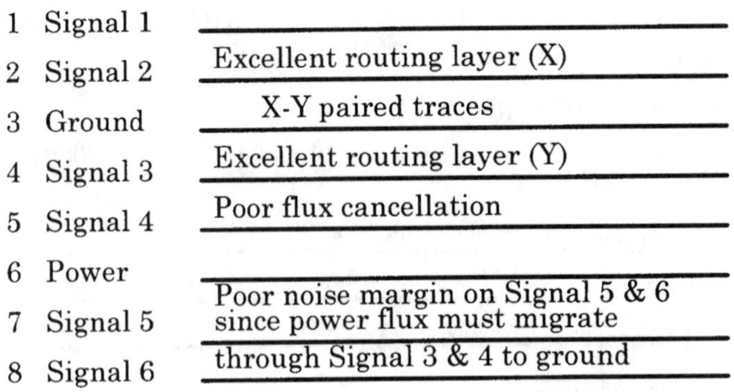

Fig. 2.9 Eight-layer PCB, configuration 1

Configuration 2. This is a preferred stackup scheme due to tight flux cancellation of RF currents. It uses four routing layers and four planes as shown in Fig. 2.10.

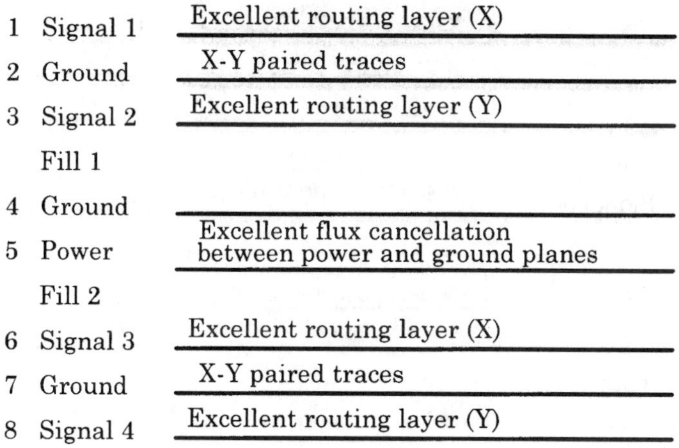

1	Signal 1	Excellent routing layer (X)
2	Ground	X-Y paired traces
3	Signal 2	Excellent routing layer (Y)
	Fill 1	
4	Ground	Excellent flux cancellation between power and ground planes
5	Power	
	Fill 2	
6	Signal 3	Excellent routing layer (X)
7	Ground	X-Y paired traces
8	Signal 4	Excellent routing layer (Y)

Fig. 2.10 Eight-layer PCB, configuration 2

- first layer (component side), microstrip signal routing layer
- second layer, ground plane
- third layer, stripline signal routing layer
- fourth layer, ground plane
- fifth layer, power plane
- sixth layer, stripline signal routing layer.
- seventh layer, ground plane
- eighth layer (solder side), microstrip signal routing layer

2.1.5 Ten-layer boards

For ten-layer boards, six routing layers are used, with four planes as shown in Fig. 2.11.

- first layer (component side), microstrip signal routing layer
- second layer, ground plane
- third layer, stripline signal routing layer
- fourth layer, stripline signal routing layer
- fifth layer, ground plane
- sixth layer, power plane
- seventh layer, stripline signal routing layer
- eighth layer, stripline signal routing layer
- ninth layer, ground plane
- tenth layer (signal), microstrip signal routing layer

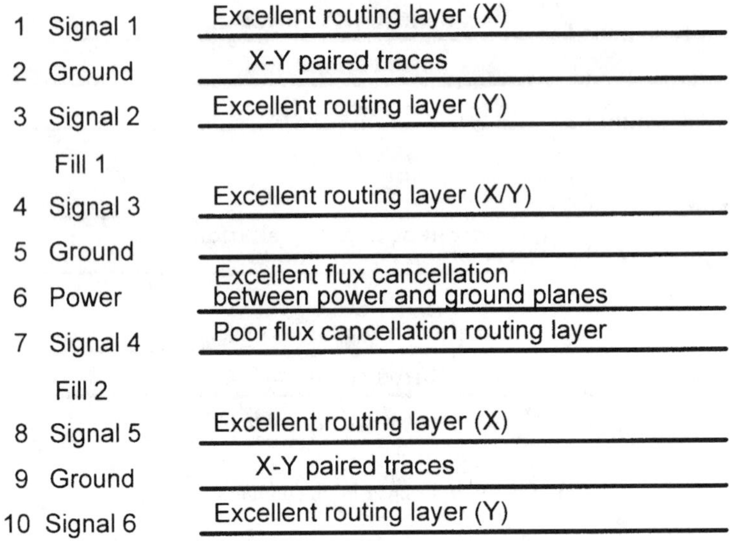

Layer	
1 Signal 1	Excellent routing layer (X)
2 Ground	X-Y paired traces
3 Signal 2	Excellent routing layer (Y)
Fill 1	
4 Signal 3	Excellent routing layer (X/Y)
5 Ground	
6 Power	Excellent flux cancellation between power and ground planes
7 Signal 4	Poor flux cancellation routing layer
Fill 2	
8 Signal 5	Excellent routing layer (X)
9 Ground	X-Y paired traces
10 Signal 6	Excellent routing layer (Y)

Fig. 2.11 Ten-layer PCB stackup

2.2 20-H RULE

RF currents exist on the edges of power planes due to magnetic flux linkage. This interplane coupling is called *fringing,* and it is generally observed only on very high-speed PCBs. When using high-speed logic and clocks, power planes can couple RF currents to each other and thus radiate RF energy into free space. To minimize this coupling effect, all power voltage planes must be physically smaller than the closest ground plane per the 20-H rule.* Figure 2.12 shows the effects of RF fringing from the edge of the PCB.

Use of the 20-H rule increases the intrinsic self-resonant frequency of the PCB. Power distribution threshold of effect is found at approximately 10-H, with 20-H representing the approximately 70 percent flux boundary. To achieve the 98 percent flux boundary, use 100-H.

To implement the 20-H Rule, determine the physical distance spacing between the power plane and its nearest ground plane. This distance spacing includes the thickness of the core, prepreg filler, and isolation separation as specified in the PCB fabrication drawing. Assuming a distance separation of 0.006 inches between planes, calculate "H" as 20 × 0.006 inches = 0.120 inches. Make the power plane 0.120 inches smaller

* The "20-H" rule was defined by W. Michael King.

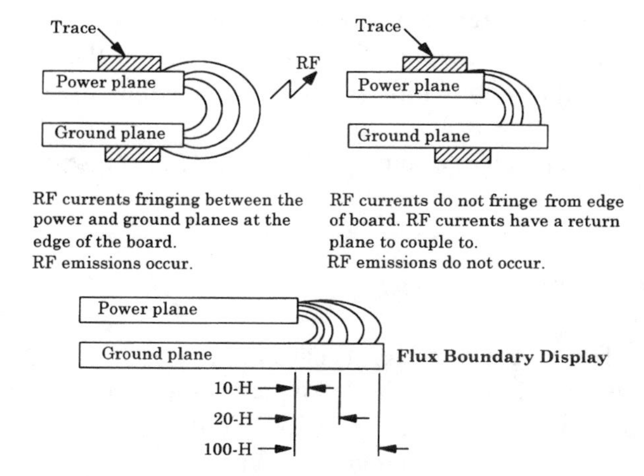

Fig. 2.12 RF fringing

than the ground plane. Should a power pin to a component be located inside this isolated (absence of copper) area, the power plane may be jogged to provide power to this isolated power pin. This is shown in Fig. 2.13.

When using the 20-H Rule, any traces on the adjacent signal routing plane located over the absence of copper area must be rerouted to be physically adjacent to a solid plane (voltage or ground), with no exceptions. It becomes important now to check two or more planes for proper implementation; i.e., 20-H and routing of traces over a solid plane.

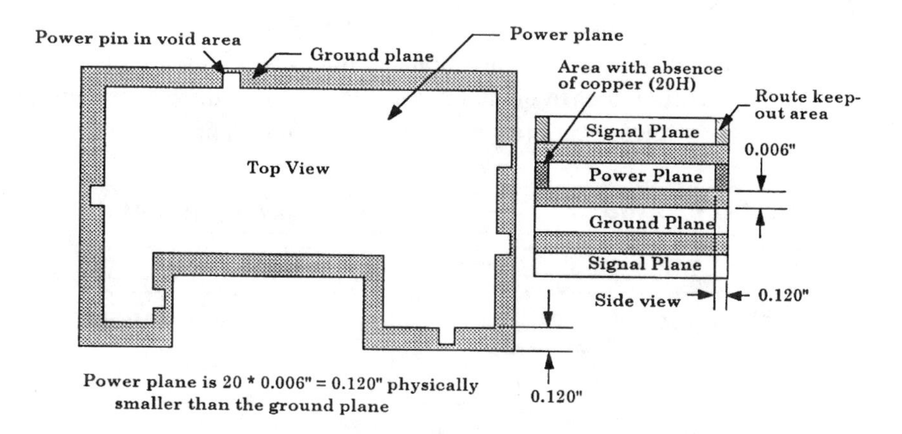

Fig. 2.13 Implementing the 20-H rule

If functional subsystem partitioning is implemented on the PCB, as discussed later in this chapter, then also implement the 20-H rule in the high-frequency bandwidth areas (CPU section, Ethernet, SCSI, etc.). When providing isolation and filtering between digital and analog sections or equivalent circuits, the 20-H rule is applicable as shown in Fig. 2.14.

2.3 GROUNDING METHODS

Many grounding schemes and terms have been devised, including digital, analog, safety, signal, noisy, quiet, earth, single-point, multipoint, and so on. Grounding methods must be specified and designed into a product—not left to chance. Designing a good grounding system is also cost-effective in the long run. In any PCB, a choice must be made between two basic types of grounding: single and multipoint. Interactions with other grounding methods can exist if planned for in advance. The proper choice of grounding is product application dependent. It must be remembered that if single-point grounding is used, the designer must be consistent in its application. The same rule exists for multipoint grounding: do not mix a multipoint ground with single-point ground unless the design allows for isolation between planes and functional subsections!

Figure 2.15 illustrates three grounding methods. The following text presents a detailed explanation of each concept as related to operational frequency and appropriate use.

2.3.1 Single-point grounding

Single-point grounding is best when the speed of components, circuits, interconnects, and the like are in the range of 1 MHz or less. At higher frequencies, the inductance of the interconnect traces will increase the PCB impedance. At still higher frequencies, the impedance of the power planes and interconnect traces become noticeable. These impedances can be very

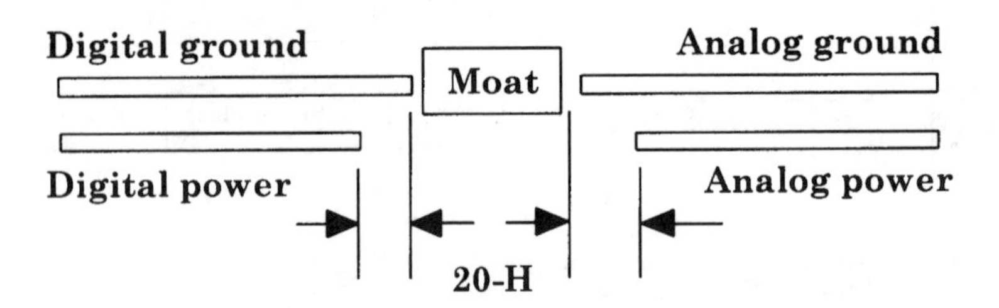

Fig. 2.14 Application of the 20-H rule and power plane isolation

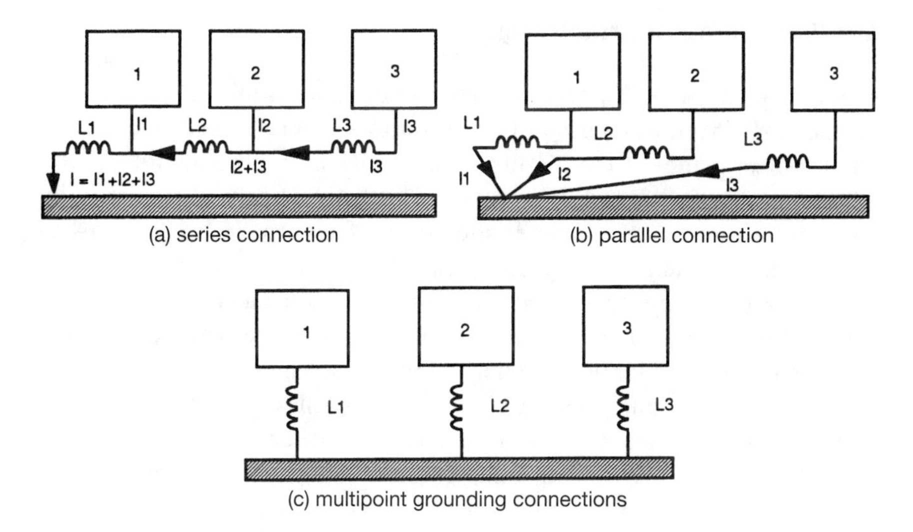

Fig. 2.15 Three different grounding methods

high if the trace lengths coincide with odd multiples of a quarter-wavelength of the edge rate of components with periodic signals. Not only will these traces and ground conductors have large impedances, they can also act as antennas and radiate RF energy. At frequencies above 1 MHz, a single-point ground generally is not used. However, exceptions do exist if the design engineer recognizes the pitfalls and designs the product using highly specialized and advanced grounding techniques.

Single-point grounds are usually formed with signal radials and are found in audio circuits, analog instrumentation, 60 Hz and dc power systems, and products packaged in plastic enclosures. Although single-point grounding is commonly used for low-frequency products, it is occasionally found in extremely high-frequency circuits and systems.

Use of single-point grounding on a CPU-motherboard or adapter (daughter) card allows for loop currents to exist between the ground planes and chassis housing if metal is used as the chassis. Loop currents create magnetic fields. These magnetic fields create electric fields and these generate RF currents. It is nearly impossible to implement single-point grounding in personal computers and similar devices, because the different subassemblies and peripherals are grounded directly to the metal chassis in different locations. These create a distributed transfer impedance between the chassis and the PCB that inherently develops loop structures. Multipoint grounding places these loops in regions where they are least likely to cause problems (i.e., they can be controlled and directed rather than allowed to transfer energy inadvertently).

2.3.2 Multipoint grounding

High-frequency designs generally require use of multiple chassis ground connections. Multipoint grounding minimizes ground impedance present in the power planes of the PCB by shunting RF currents from the ground planes to chassis ground. Low plane impedance is caused primarily by the lower inductance characteristic of solid planes. In very high-frequency circuits, lengths of ground leads from components must also be kept as short as possible. Trace lengths add inductance to a circuit—approximately 15–20 nH per inch, depending on trace width and height above a plane. This inductance may allow a resonance to occur when the distributed capacitance between the ground planes and chassis ground form a tuned resonant circuit. The capacitance value, C, in Eq. (2.1) can be determined as shown in Chapter 3. Inductance, L, is determined through knowledge of the impedance of the copper planes, as discussed in Chapter 4.

$$f = \frac{1}{2\pi\sqrt{LC}} \tag{2.1}$$

where

f = resonant frequency (Hz)

L = inductance of the circuit (Henrys)

C = capacitance of the circuit (Farads)

Equation (2.1) describes most aspects of frequency-domain concerns. This equation, although simple in format, requires knowledge of how to calculate both L and C, which by themselves are not easy to determine intuitively. Chapters 3 and 4 makes this equation easy to use and implement.

Examine Eq. (2.1) using Fig. 2.16. This illustration shows the capacitance and inductance that exist in a PCB and mounting panel. Capacitance and inductance will always exist. Depending on the aspect ratio (between the mounting posts) in relation to the self-resonant frequency of the power planes, loop currents will be generated and coupled (either radiated or conducted) to other PCBs located adjacent to each other, the chassis housing, internal cables or harnesses, peripheral devices, I/O circuits and connectors, or to free space.

In addition to inductance in the planes, long traces also act as small antennas, especially for clock signals and other periodic data pulses. By minimizing trace inductance and removing RF currents created in the trace (coupling RF currents present in the signal trace to the ground

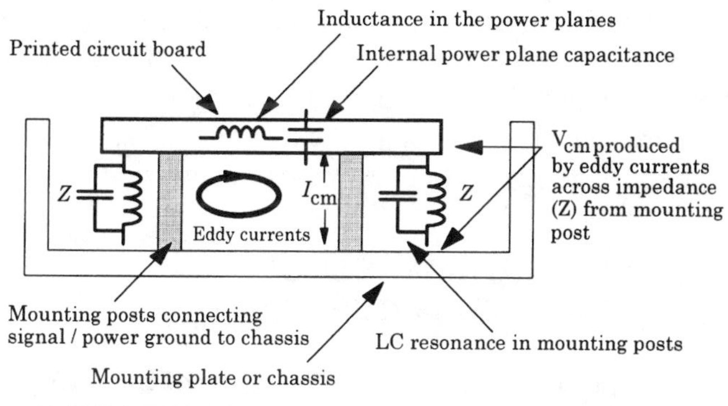

APPLICATION MODEL OF MULTI-POINT GROUNDING

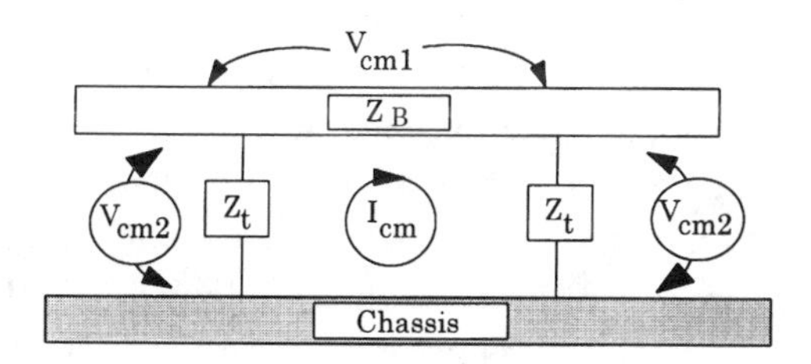

V_{cm2} is reduced by the mounting posts shown above.

With reduction of V_{cm2} we have a reduction in EMI between the board and chassis.

Resonance is thus removed along with enhanced RF suppression.

Z_t is the distributed transfer impedance.

ELECTROMAGNETIC MODEL PRIOR TO MULTI-POINT GROUNDING

Fig. 2.16 Resonance in distributive field transfers

planes or chassis ground), significant improvement in signal quality and RF suppression will occur. The subject of trace impedance and inductance is discussed in detail in Chapter 4.

Digital circuits must be treated as high-frequency analog circuits. A good low-inductance ground is necessary on any PCB containing many logic circuits. The ground planes internal to the PCB (more than the power plane) generally provide a low-inductance ground return for the power supply and signal currents. This allows for the possibility of using a constant impedance transmission line for signal interconnects. When making ground plane(s) to chassis plane connection, provide for high-frequency decoupling of RF current. These currents are created by the resonant circuit of the planes and their relationship to signal traces. Use high-quality bypass capacitors, usually 0.1 μF in parallel with 0.001 μF on voltage planes at each and every ground connection, as will be reiterated in Chapter 3. The chassis grounds are frequently connected directly to the ground planes of the PCB to minimize RF voltages and currents that exist between board and chassis. If magnetic loops are small (1/20 wavelength of the highest RF generated frequency), RF suppression is enhanced.

2.4 GROUND AND SIGNAL LOOPS (EXCLUDING EDDY CURRENTS)

Loops are a major contributor to the propagation of RF energy. RF current will attempt to return to its source through any available path or medium: components, wire harnesses, ground planes, adjacent traces, and so forth. RF current is created between a source and load in the return path. This is due to the voltage potential difference between two ground sources, regardless of whether inductance exists between these loads. Inductance, however, causes magnetic coupling of RF current to occur between a source and victim circuit, and increases RF losses in the path.

One of the most important design considerations for EMI suppression on a PCB is ground or signal return loop control. An analysis must be made for each and every ground stitch connection (mechanical securement between the PCB ground and chassis ground) related to RF currents generated from RF noisy electrical circuits. Always locate high-speed logic components and oscillators as close as possible to a ground stitch connection to minimize RF loops in the form of eddy currents to chassis ground.

An example of loops that could occur in a computer with adapter cards and single-point grounding is shown in Fig. 2.17. As observed, an excessive signal return loop area exists. Each loop will radiate a distinct electromagnetic field. RF currents will create an electromagnetic field at a unique frequency. Containment measures must now be used to keep these RF currents from coupling to other circuits or radiating to the external environment as EMI. Internally generated RF loop currents are to be avoided.

For example, $\lambda/20$ of a 64 MHz oscillator is 23 cm (9.2 inches). If the straight-line distance between any two ground screws (in both x and y axes) is greater than 9.2 inches, then a potential efficient RF loop exists. This loop could be the source of RF energy propagation, which could cause noncompliance with international EMI emission limits. Unless other design measures are implemented, suppression of RF currents caused by poor loop control is not possible, and containment measures (i.e., sheet metal) must be implemented. Sheet metal is an expensive band-aid that is not always effective for RF containment. An example of *aspect ratio* is illustrated in Fig. 2.18.

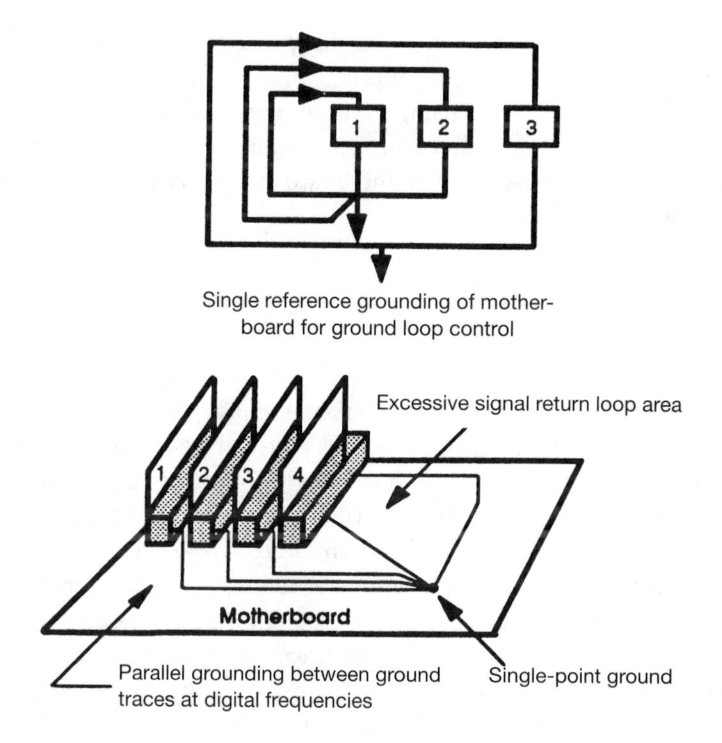

Fig. 2.17 Ground loop control (Source: CKC Laboratories, reprinted by permission)

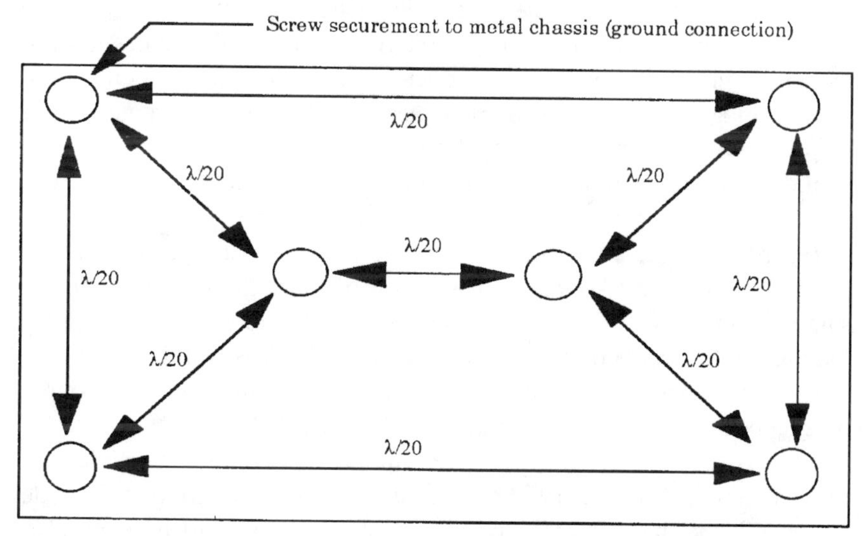

Fig. 2.18 Aspect ratio on a PCB

RF currents in power planes also have the tendency to couple, via crosstalk, to other signal traces, thus causing improper operation or functional signal degradation. If using multipoint grounding, consideration of loops becomes a major design concern. (Note again the multiple ground and signal return loops in a motherboard configuration with daughter cards shown in Fig. 2.17.)

2.5 IMAGE PLANES

An image plane is a layer of copper (voltage plane, ground plane, chassis plane, or isolated plane) internal to a PCB physically adjacent to a circuit or signal plane. Use image planes to provide a low impedance path for RF signal currents to return to their source (flux return), thus completing the RF current return path and reducing EMI emissions. The term *image plane* was popularized by Ref. [2] and is now used as industry standard terminology.

RF currents must return to their source one way or another. This path may be a mirror image of its original trace route, through another trace located in the near vicinity (crosstalk), a power plane, a ground plane, or chassis plane. RF currents will capacitively (or by mutual inductance) couple themselves to a conductive medium (i.e., low-impedance path such as the copper that makes up a plane). If this coupling is not 100 percent, common-mode RF current can be generated between traces and their nearest plane. An image plane internal to the PCB reduces ground noise voltage in addition to allowing RF currents to return to their source (mirror image) in a tightly coupled (nearly 100 percent) manner. Tight coupling provides for flux cancellation, which is another reason for use of a solid plane. Solid planes also prevent common-mode RF current from being generated in the PCB by traces rich in RF energy [2].

Regarding image plane theory, the material presented in this guideline is based upon a finite-sized plane, typical of most PCBs. Image planes cannot be relied upon for reducing currents on I/O cables because approximating finite-sized conductive planes is not always valid. When I/O cables are provided, the dimensions of the configuration and source impedance are important parameters to remember [3].

An illustrative example of common-mode and differential-mode currents is shown in Fig. 2.19. The measured **E**-field of differential-mode current will be the difference of I1 and I2. This sum will be negligible because of a 180° phase difference. The measured **E**-field due to common-mode current will be the sum of I1 and I2, which could be substantial due to the summing effect.

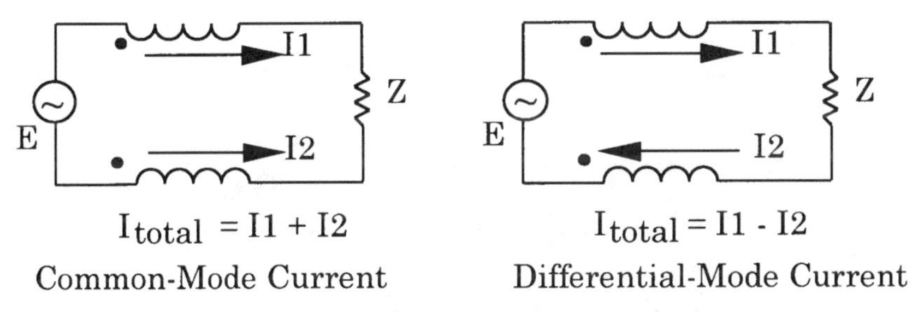

Fig. 2.19 Common-mode and differential-mode currents

If three internal signal planes are physically adjacent in a multilayer board, then the middle signal plane (i.e., the one not adjacent to an image plane) will couple its RF currents to the other two signal planes, thus causing RF energy to be transferred (by mutual inductance and capacitive coupling) to the other two signal planes. This coupling can cause significant crosstalk to occur, which may include non-functionality. Flux cancellation performance is enhanced when the signal routing layer is adjacent to a ground plane and not adjacent to a power plane, as described earlier in this chapter.

For an image plane to be effective, *no signal traces can be located in this solid plane.* Exceptions exist when a moat (or isolation) occurs. If a signal trace, or even a power trace (i.e., +12 V trace in a +5 V plane) is routed in a solid plane, this plane is now fragmented into smaller parts. Provisions have now been made for a ground or signal return loop to exist for signal traces that are routed on an adjacent layer across this violation. This loop occurs by not allowing RF currents in a signal trace to seek a straight line path back to its source.

Figure 2.20 illustrates a violation of the image plane concept. These planes can now no longer function as a solid plane to remove common-mode RF currents, and the losses across the plane segmentations may actually produce RF fields. Vias placed in an image plane do not degrade the imaging capabilities of the plane, except where ground slots are provided, as discussed next.

Another area of concern that lies with ground plane discontinuities is the use of through-hole components. Excessive use of through-holes in a power or ground plane creates the "Swiss cheese syndrome" [4]. The copper area in the plane is reduced because many holes overlap (oversized through-holes), leaving large areas of discontinuity. This is observed in Fig. 2.21. The return current flows on the image plane, while the signal trace may be a direct line route on its separate plane across the discontinuity. As seen in Fig. 2.21, the return currents in the ground plane must travel around slots or holes. As a result, extra trace length exists for return

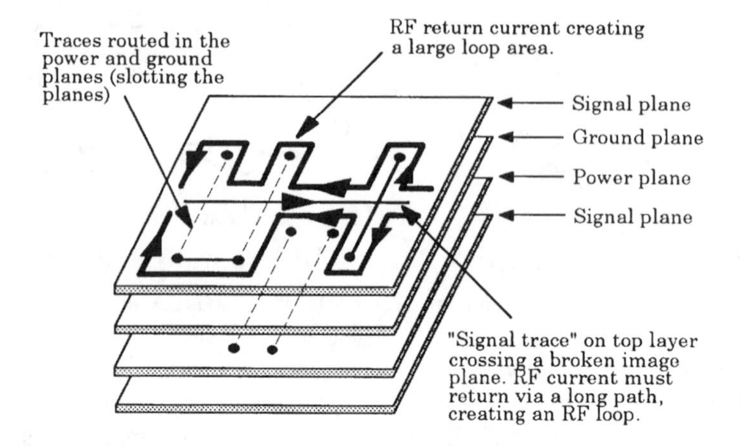

Fig. 2.20 Four-layer stackup

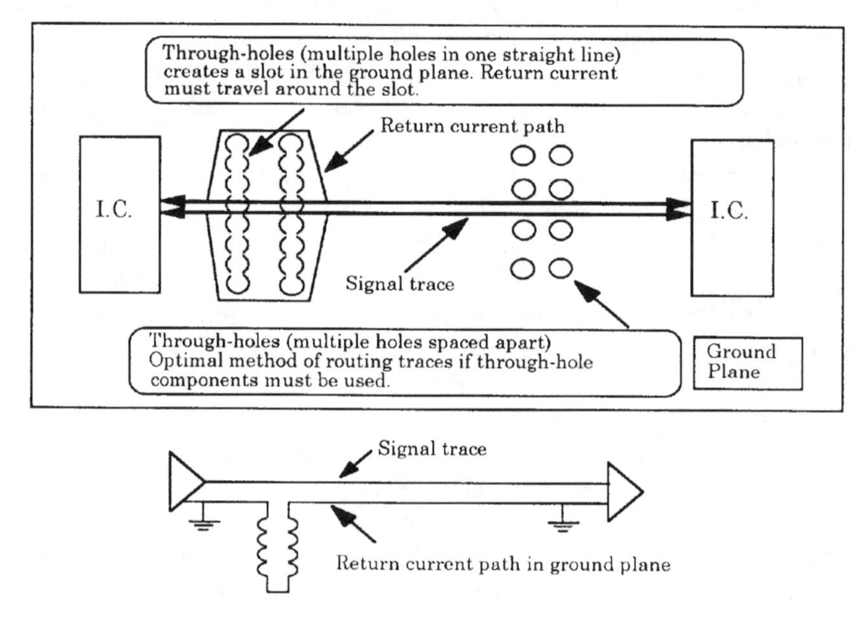

Fig. 2.21 Ground slots (loops) when using through-hole components

currents that must flow around these slots in the image plane. This extra length adds additional inductance into the signal return trace. With additional inductance in the return path, there exists reduced common-mode coupling between the signal trace and RF current return plane (less flux cancellation). For through-hole components that have a space between pins (non-oversized holes), optimal reduction of signal and return current is achieved due to less inductance in the signal return path and the existence of the solid plane.

If the signal trace is routed "around" through-hole discontinuities (left-hand side of Fig. 2.21), maintain a constant image plane along the entire signal route. The same is true for the right-hand side of Fig. 2.21. There exists no ground plane discontinuities and, hence, shorter trace length. The longer trace route on the left side in Fig. 2.21 adds additional trace length inductance. This length may cause reflections that affect signal quality and functionality. Problems arise when the signal trace travels through the middle of slotted holes in the PCB (in an attempt to minimize trace length routing) and when a solid plane does not exist in this oversized through-hole area. When routing traces between through-hole components, use of the 3-W rule* must be maintained between the trace and through-hole clearance area.

Generally, a slot in a PCB with through-hole components will not cause RF problems for the majority of signal traces that are routed between through-hole device leads. For high-speed, high-threat signals, alternative methods of routing traces between through-hole component leads must be devised.

In addition to reducing ground-noise voltage, image planes prevent ground loops from occurring because RF currents tightly couple to their source without having to find another path home. Loop control is maintained and minimized, and flux cancellation is optimized. This is one of the most important concepts of EMI suppression at the PCB level. Proper placement of an image plane adjacent to each and every signal plane removes common-mode RF currents created by signal traces. Image planes carry high levels of RF currents that must be sourced to ground potential. To help remove this excess RF current, all ground and chassis planes must be connected to chassis ground via a low-impedance ground stitch connection [2,5,6]. Figure 2.22 illustrates the concept and use of an image plane.

There is one concern related to image planes. This deals with the concept of skin effect. *Skin effect* refers to current flow that resides in the first skin depth of the material. Current does not and cannot significantly flow in the center of traces and wires—and is predominantly observed on the outer surface of the conductive media. Different materials have different skin depth values. The skin depth of copper is extremely shallow above 30 MHz. Typically, this is observed at 0.00026 inch (0.0066 mm) at 100 MHz. RF current present on a ground plane cannot penetrate 0.0014 inch

* Described by W. Michael King

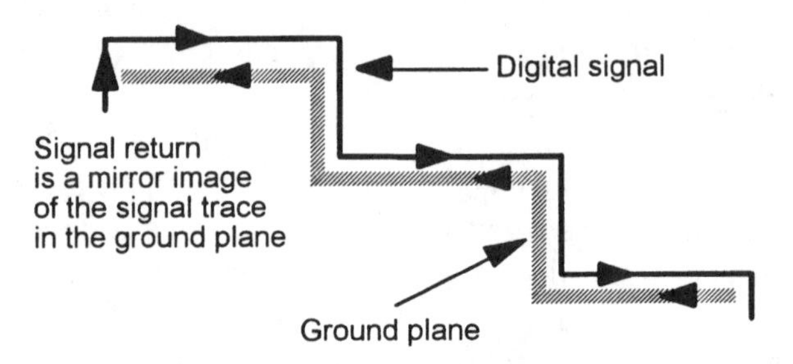

Fig. 2.22 Image plane concept

(0.036 mm) thick copper. As a result, both common-mode and differential-mode currents flow only on the top (skin) layer of the plane. There is no significant current flowing internal to the image plane or on its bottom. Placing an additional image plane beneath this ground plane would not provide additional EMI reduction. If the second plane is at voltage potential (the primary plane at ground potential), then a decoupling capacitor would be created. These two planes can now be used as both a decoupling capacitor and dual image planes [7], but with some concern regarding flux cancellation (see section 2.1).

2.6 PARTITIONING

Proper placement of components is crucial in any PCB layout. Most designs incorporate functional subsections or areas (by logical function). Grouping each functional area adjacent to other subsections minimizes signal trace lengths and reflections, makes trace routing easier, and maintains signal quality. Avoid the use of vias where possible. Vias increase the inductance of the trace by approximately 1 to 3 nH per via. Figure 2.23 illustrates functional grouping of subsections (or areas) on a stand-alone CPU motherboard.

Extensive use of chassis ground stitch connections is also observed in Fig. 2.23. High-frequency designs require new methodologies for bonding ground planes to chassis ground. Use of these multipoint grounding techniques effectively partitions common-mode eddy currents emanating from various segments in the design and keeps them from coupling into other segments. Products with clocks above 50 MHz generally require frequent ground stitch connections to chassis ground to minimize effects of common-mode eddy currents and ground loops present between func-

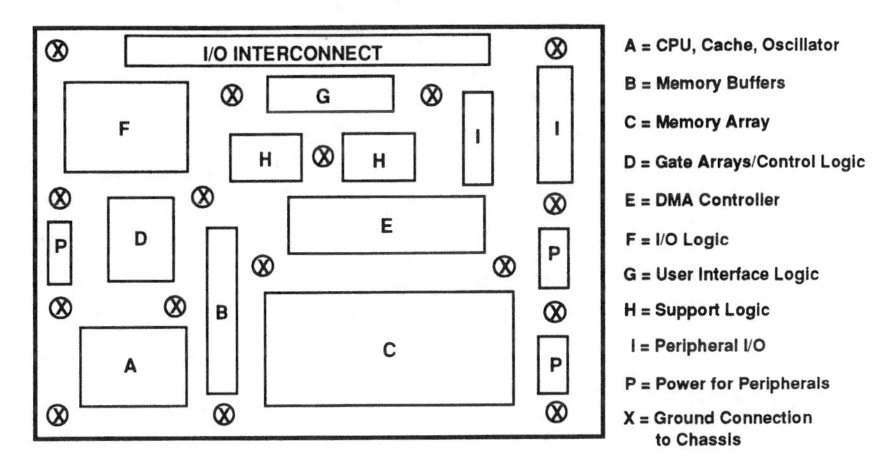

Fig. 2.23 Partitioning with multipoint grounding

tional sections. At least four ground points surround each subsection. These ground points illustrate best-case implementation of aspect ratio. Note that a chassis bond connection (screw or equivalent) is located on both ends of the dc power connector (Item P) used for powering external peripheral devices. RF noise generated on either the PCB or peripheral power subsystem must be ac shunted to chassis ground by parallel bypass capacitors. These capacitors reduce coupling of power-supply-generated RF currents into signal and data lines. Removal of RF currents on the power connector will optimize signal quality for data transfer between the motherboard and external peripheral devices, in addition to reducing emissions [5].

Most PCBs consist of functional subsections or areas. A typical CPU motherboard contains the following: CPU, memory, ASICs, I/O, bus interface, system controllers, EISA/ISA bus, SCSI bus, peripheral interface (fixed and floppy disk drives), and other components. Associated with each subsection are different bandwidths of RF energy. Different logic families generate RF energy at different portions of the frequency spectrum. The higher the frequency component of the signal, the greater the bandwidth of RF spectral energy. RF energy is generated from the higher frequency components and the time variant edges of digital and analog signals. Clock signals are the greatest contributors to the generation of RF energy. This is because clock signals are periodic (50 percent duty cycle) and are easy to measure with spectrum analyzers or receivers. Table 2.2 illustrates the spectral energy bandwidth of various logic families.

To prevent coupling between different bandwidth areas, functional partitioning is used. *Partitioning* refers to the physical separation between

Table 2.2 Chart of Logic Families

Logic Family	*Rise/Fall Time (Approx.)* T_r	*Principal Harmonic Content* $F_2 = (1/\pi T_r)$	*Possible Significant Spectrum* $F_{max} = 10 \times F_2$	*Maximum Non-transmission Line Trace Length (Microstrip)* $L_{max} = 9 \times T_r$	*Maximum Non-transmission Line Trace Length (Stripline)* $L_{max} = 7 \times T_r$
74L xxx	31–35 ns	10 MHz	100 MHz	279 cm (110")	217 cm (85.4")
74C xxx	25–60 ns	13 MHz	130 MHz	225 cm (88.5")	175 cm (69")
CD4 xxx (CMOS)	25 ns	13 MHz	130 MHz	225 cm (88.5")	175 cm (69")
74HC xxx	13–15 ns	24 MHz	240 MHz	117 cm (46")	91 cm (36")
74 xxx (flip-flop)	10–12 ns 15–22 ns	32 MHz 21 MHz	320 MHz 210 MHz	90 cm (35.5") 135 cm (53")	70 cm (27.5") 105 cm (41")
74LS xxx (flip-flop)	9.5 ns 13–15 ns	34 MHz 24 MHz	340 MHz 240 MHz	85.5 cm (34") 117 cm (46")	66.5 cm (26") 91 cm (36")
74H xxx	4–6 ns	80 MHz	800 MHz	36 cm (14.2")	28 cm (11")
74S xxx	3–4 ns	106 MHz	1.1 GHz	27 cm (10.5")	21 cm (4.3")
74HCT xxx	5–15 ns	64 MHz	640 MHz	45 cm (18")	35 cm (14")
74ALS xxx	2–10 ns	160 MHz	1.60 GHz	18 cm (7")	10 cm (4")
74ACT xxx	2–5 ns	160 MHz	1.60 GHz	18 cm (7")	10 cm (4")
74F xxx	1.5–1.6 ns	212 MHz	2.1 GHz	10.5 cm (4")	10.5 cm (4")
ECL 10K	1.5 ns	212 MHz	2.1 GHz	10.5 cm (4")	10.5 cm (4")
ECL 100K	0.75 ns	424 MHz	4.2 GHz	6 cm (3")	5.25 cm (2")

T_r depends greatly on load capacitance, supply voltage, and IC complexity. Consult manufacturer's specifications.
Note: T_r will differ between device manufacturers and their fabrication processes.

Assume 1.7 ns/ft (0.14 ns/inch or 0.36 ns/cm) propagation delay for FR-4, $E_r = 4.6$ (microstrip).
Assume 2.2 ns/ft (0.18 ns/inch or 0.47 ns/cm) propagation delay for FR-4, $E_r = 4.6$ (stripline).

functional sections. Partitioning is product specific and may be achieved using separate PCBs, isolation, topology layout variations, or by other creative means. Partitioning methods are detailed in Chapter 5.

Proper partitioning allows for optimal functionality, ease of routing traces, and minimized trace lengths. It allows smaller loops to exist while optimizing signal quality. The design engineer will specify which components are associated with each functional subsection. Use the information provided by the component manufacturer and design engineer to optimize component placement prior to routing any traces.

2.7 LOGIC FAMILIES

When selecting a digital logic device for a particular application, design engineers are generally interested in only functionality and operating speed. The speed of the component is based on the propagation delay of the internal logic gates as published by the manufacturer.

As components become faster (faster internal propagation), increases in RF return currents, crosstalk, and ringing will occur, based on the inverse relationship between speed and EMI. A device is generally chosen based on the propagation from input to output, along with the setup time of the input signal. Almost all components have internal logic gates that operate at a faster edge rate than the propagation delay required for functionality. As a result, slower logic families (internal gates) are preferred, since propagation delay is the primary function of the circuit. Figure 2.24 illustrates the relationship between the internal switching speed of a basic inverter gate compared to propagation delay.

Various logic families are available with different design features. These features vary between CMOS, TTL, and ECL. Some of these features include input power, package outline, speed-power combinations, voltage swing level, and edge rates. Certain logic devices are now available with clock skew circuitry to control the internal edges of the internal logic gates while maintaining accurate propagation delay.

One extremely important device parameter (for EMC) usually not specified by device manufacturers is *"power peak inrush surges into the device power pins."* These peak power currents are the result of logic crossover currents, device capacitive overheads, and capacitance caused by surge currents from trace capacitance and loading device junctions. These "surge" power currents may exhibit levels that are many multiples of the actual signal currents driven!

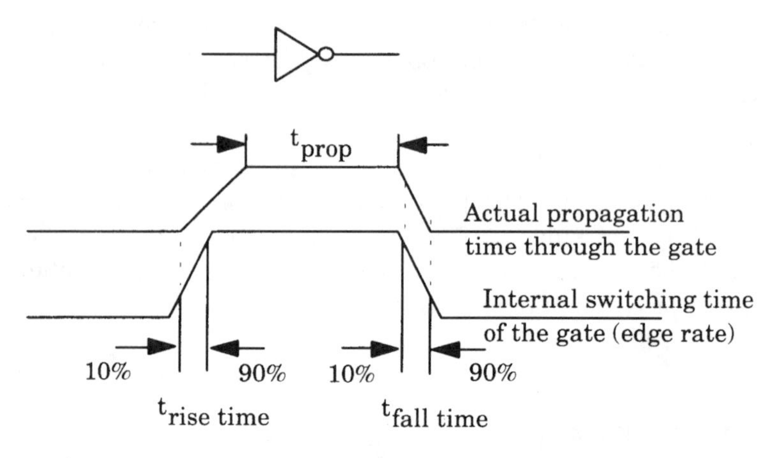

Fig. 2.24 Output switching time vs. propagation delay

Selection of the slowest possible logic family while maintaining adequate timing margins minimizes EMI effects and enhances signal quality. Use devices with edge times (t_r) greater (or slower) than 5 ns, if possible, for applications where the timing requirement of functionality allows. Recognize the fact that use of standard and low-power Schottky TTL logic (i.e., 74LS series) is becoming less common in today's marketplace. Design techniques presented in this guide are usually not required when using these slower-speed logic families. However, today's high-speed, high-technology products require use of extremely fast-edge logic in the 1.5–5 ns range; i.e., 74ACT and 74F series components. Use of a 74HCT could be provisionally adequate for replacement of a 74ACT device for some applications, with the added benefit of less RF emissions generated. As a general consideration, *do not use faster devices than the functional timing diagram of the circuit actually demands.*

If timing purposes require fast logic families, the designer must address individually the issues of special decoupling, routing, and handling of clock traces. These are discussed in Chapter 4. Refer again to Table 2.2 for details on EMI characteristics of different logic families.

Fast switching times (edges) cause a proportional increase in problems related to return currents, crosstalk, ringing, and reflections. These problems are independent of propagation delay. This is because logic families have edge rates that are faster than the propagation delay inherent in the device. No two logic families are the same. Even the same components from different manufacturers differ in construction and minimum edge rates. Edge rate is defined as the rate of voltage or current change per unit time (volts/ns or amperes/ns).

When selecting a logic family, manufacturers will specify in their data books the "maximum" edge rate, $t_{r\ max}$, of the clocks and I/O pins. Generally, this specification is usually 2–5 ns maximum. It is observed that the "minimum" edge rate, $t_{r\ min}$, is not stated. A device with a 2 ns maximum edge rate specification may in reality be a 0.5 to 1.0 ns edge device. The greatest contributor to RF energy is the edge rate of the device, not the actual operating frequency. A 5 MHz oscillator driving a 74F04 driver (with a 1 ns edge) will generate larger amounts of RF spectral energy over the frequency spectrum than a 10 MHz oscillator driving a 74ALS04 (with a 4 ns edge). *This one component specification is the most frequently overlooked and forgotten parameter in PCB design; however, this is the most critical aspect with which a design engineer must be concerned to ensure an EMI compliant product. The frequently heard statement, "Use the slowest logic family possible,"* is a result of this minimum edge rate parameter **not being specified or published** by a component manufacturer in their data books. Edge rates of digital components are the source of almost all RF energy created within a PCB.

Use of surface mount technology (SMT) components is preferred over through-hole mounting for minimizing RF emissions. This characteristic difference is due to shorter lead length (lower inductance) from the die of the SMT component to the circuit trace on the PCB. Also, SMTs have smaller "loop areas" due to the smaller package size. Trace lengths add inductance. Inductance is a component that results in unwanted RF energy. RF currents are a cause of RF emissions, in addition to possibly causing signal quality problems. Sometimes through-hole devices are installed on sockets. Sockets also add additional lead length inductance and, hence, potentially greater amounts of EMI.

The reason for using the slowest logic family possible stems from the relationship between the time domain and frequency domain. Fourier analysis of signal edges in the time domain shows that, as the slope (edge rate) of the signal becomes faster, a greater amount of spectral bandwidth of RF energy is created. A detailed discussion of Fourier transforms and analysis is beyond the scope of this guide.

If possible, select logic devices with power and ground pins located in the middle of the device rather than on opposite corners. Power pins in the center of the component provide for optimal placement of decoupling capacitors (when placed on the bottom side of the PCB). This configuration also minimizes trace length connections between the device and decoupling capacitor, in addition to minimizing trace length inductance from the power and ground pins internal to the silicon wafer (die) of the package.

Table 2.2 also provides details on the harmonic spectrum of digital logic families. Refer to this chart in determining the best logic family for minimizing EMI emissions while allowing for proper functionality of the design.

2.8 VELOCITY OF PROPAGATION

This section is provided as background for further discussion throughout this book. Velocity of propagation, V_p, is the speed at which data is transmitted through conductors or in a PCB. In air, the velocity of propagation is the speed of light. In a dielectric material, the velocity is slower and is given by

$$V_p = \frac{C}{\sqrt{\varepsilon}} \qquad (2.2)$$

where

$C = 3 \times 10^8$ meters per second, or about 12 inches/ns (30 cm/ns)

ε_r = effective dielectric constant

Typically, ε_r is about 3 or 4 for PCBs, even through the relative dielectric constant of the board material is near 5. This is because part of the energy flow is in air, and part is in the dielectric medium. A typical dielectric constant yields a velocity of propagation of 6 to 7 inches per nanosecond.

2.9 CRITICAL FREQUENCIES (λ/20)

To determine the frequency, f, of a signal and its related wavelength, λ, use the following conversion equations:

$$f(\text{MHz}) = \frac{300}{\lambda(\text{m})} = \frac{984}{\lambda(\text{ft})}$$

$$\lambda(\text{m}) = \frac{300}{f(\text{MHz})}$$

$$\lambda(\text{ft}) = \frac{984}{f(\text{MHz})} \qquad (2.3)$$

Throughout this design guide, reference is made to critical frequencies or high-threat clock and periodic signal traces that have a length greater than λ/20. A summary of miscellaneous frequencies and their respective wavelength distances is given in Table 2.3, based on the equations shown above.

Table 2.3 λ/20 Wavelength at Various Frequencies

Frequency of Interest (MHz)	*λ/20 Wavelength Distance*
10	1.5 m (5 ft)
27	0.56 m (1.8 ft)
35	0.43 m (1.4 ft)
50	0.3 m (12 in)
80	0.19 m (7.52 in)
100	0.15 m (6 in)
160	9.4 cm (3.7 in)
200	7.5 cm (3 in)
400	3.6 cm (1.5 in)
600	2.5 cm (1.0 in)
1000	1.5 cm (0.6 in)

2.10 REFERENCES

1. Technical definition and explanation provided by W. Michael King.
2. German, R.F., H. Ott, and C.R. Paul. 1990. Effect of an image plane on PCB radiation. *Proceedings of the IEEE International Symposium on Electromagnetic Compatibility.* New York: IEEE, 284–291.
3. Hsu, T. 1991. The validity of using image plane theory to predict PCB radiation. *Proceedings of the IEEE International Symposium on Electromagnetic Compatibility.* New York: IEEE, 58–60.
4. Mardiguian, M. 1992. *Controlling Radiated Emissions by Design.* New York: Van Nostrand Reinhold.
5. Montrose, M.I. 1991. Overview of design techniques for printed circuit board layout used in high technology products. *Proceedings of the IEEE International Symposium on Electromagnetic Compatibility.* New York: IEEE.
6. 1991. *EMI Considerations for High Speed System Design.* CKC Laboratories.
7. Dockey, R.W., and R.F. German. 1993. New techniques for reducing printed circuit board common-mode radiation. *Proceedings of the IEEE International Symposium on Electromagnetic Compatibility.* New York: IEEE, 334–339.

3

Bypassing and Decoupling

Bypassing and *decoupling* are techniques for preventing energy transference from one circuit to another. Three circuit areas are of primary concern: power and ground planes, components, and internal power connections.

Decoupling is required to provide sufficient dc voltage and current for proper operation of components during clock or data transitions when all component signal pins switch simultaneously under maximum capacitive load. Decoupling is accomplished by ensuring that there is a low-impedance power source present in both circuit traces and power planes. Because decoupling capacitors have an increasingly low impedance at high frequencies, up to the point of self-resonance, high-frequency noise is effectively removed from the signal trace, while low-frequency RF energy remains relatively unaffected. Optimal implementation is achieved by using bulk, bypass, and decoupling capacitors. All capacitor values must be calculated for each specific function. In addition, we must properly select the dielectric material of the capacitor; it cannot be left to random choice based on past experience. The following defines three common uses of capacitors for EMI reduction. Of course, a capacitor may also be used in other applications, such as timing components, wave shaping, integration, filtering, and so on.

Decoupling. Decoupling removes RF energy generated on the power planes by high-frequency components. Decoupling capacitors also provide a localized source of dc power for devices or components, and are particularly useful in reducing peak current surges propagated across the circuit board.

Bypassing. Bypassing removes unwanted RF noise that couples component or cable common-mode EMI into susceptible areas and provides other functions of filtering (bandwidth limiting).

Bulk. Bulk capacitors are used to maintain constant dc voltage and current to components when all signal pins switch simultaneously under maximum capacitive load. It also prevents power dropout due to dI/dt current surges generated by components.

3.1 RESONANCE

A capacitor, in reality, consists of an LCR circuit where L = inductance related to lead length, R = resistance in the leads, and C = capacitance. A schematic representation of a capacitor is shown in Fig. 3.1. At a calculable frequency, the series combination of L and C becomes resonant, providing very low impedance and effective RF shunting at resonance. At frequencies above self-resonance, the impedance of the capacitor becomes increasingly inductive, and bypassing or decoupling becomes less effective. Hence, bypassing and decoupling performance is affected by lead lengths of capacitors (including surface mount, radial, and axial styles), the trace length between active components and the capacitor, feed-through pads, and so forth.

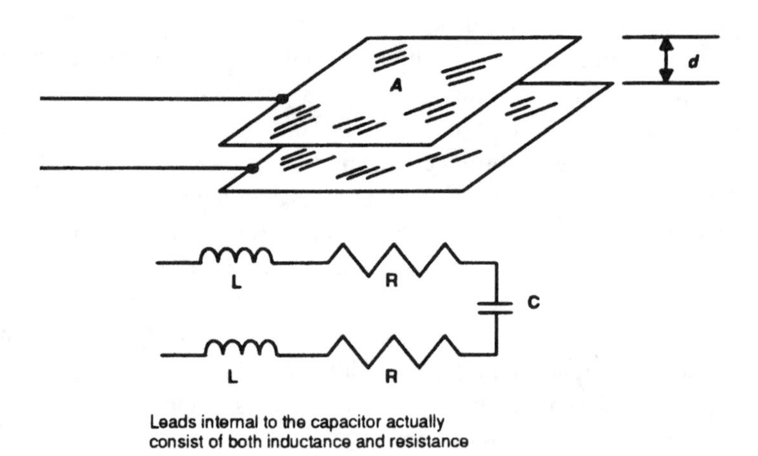

Fig. 3.1 Physical characteristics of a capacitor

Before discussing bypassing and decoupling of circuits on a PCB, a review of resonance is provided.

Resonance occurs in a circuit when the phase angle difference between the inductive and capacitive vectors is zero. This is equivalent to saying that the circuit is purely resistive in its response to ac voltage. Three types of resonance are common:

- series resonance
- parallel resonance
- parallel C–series RL resonance

Resonant circuits are frequency selective since they pass more (or less) RF current at certain frequencies than at others. A series LCR circuit will pass the selected frequency (as measured across C) if R is high and the source resistance is low. If R is low and the source resistance is high, the circuit will reject the chosen frequency. A parallel resonant circuit placed in series with the load will reject the chosen frequency.

Series resonance

The overall impedance in rectangular form of a series RLC circuit, shown in Fig. 3.2, is $R + j(X_L - X_c)$. If an RLC circuit is to behave resistively, then the values can be calculated as shown in Fig. 3.2, where ω is known as the *resonant frequency.*

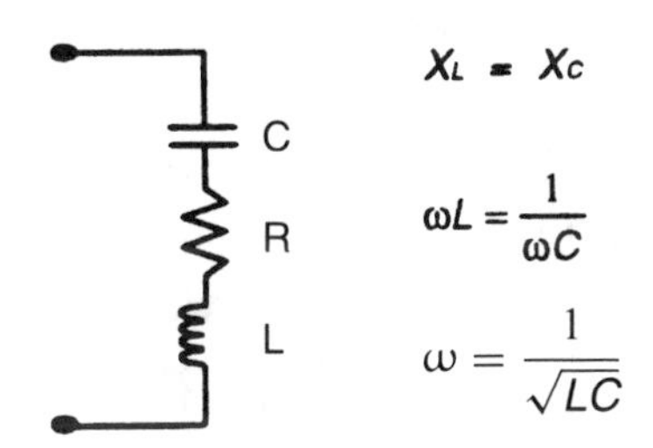

Fig. 3.2 Series resonance

With a series RLC circuit at resonance,

- Impedance is at minimum.
- Impedance equals resistance.
- The phase angle difference is zero.
- Current is at maximum.
- Power is at maximum.

Parallel resonance

A parallel LCR circuit behaves as shown in Fig. 3.3. The resonant frequency is the same as for a series LCR circuit.
With a parallel RLC circuit at resonance,

- Impedance is at maximum.
- Impedance equals resistance.

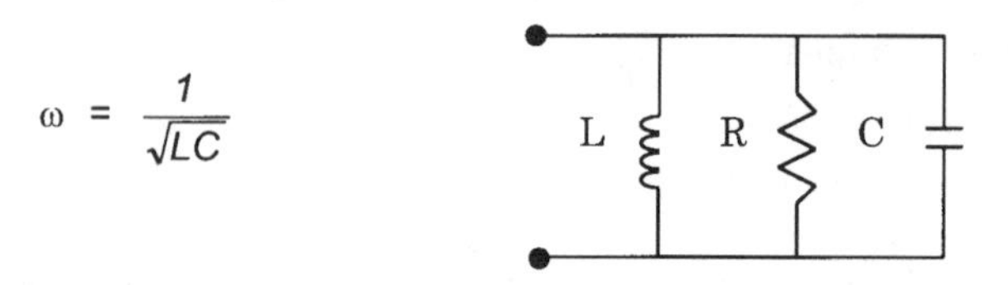

Fig. 3.3 Parallel resonance

- The phase angle difference is zero.
- Current is at minimum.
- Power is at minimum.

Parallel C–series RL resonance

Practical resonant circuits generally consist of an inductor and variable capacitor in parallel. Since the inductor will possess some resistance, the equivalent circuit is as shown in Fig. 3.4. With this form of resonance, the current will be minimized at resonance.

$$\omega = \sqrt{\frac{1}{LC} - \left(\frac{R}{L}\right)^2}$$

C
R
L

Fig. 3.4 Parallel C–series RL resonance

3.2 CAPACITOR PHYSICAL CHARACTERISTICS

Decoupling capacitors ideally should be able to supply all the current necessary during a state transition of a logic device. This is described by Eq. (3.1). Use decoupling capacitors on two-layer boards to reduce power supply ripple. Decoupling capacitors usually are not needed for low-frequency applications when multilayer boards are used, given that the capacitance between the power plane and ground planes provides overall decoupling.

$$C = \frac{\Delta I}{\Delta V / \Delta t} = \frac{20\ \text{mA}}{100\ \text{mV}/5\ \text{ns}} = 0.001\ \mu\text{F or } 1000\ \text{pF} \tag{3.1}$$

where

ΔI = current transient

ΔV = allowable power supply voltage change (ripple)

Δt = switching time

When selecting bypass and decoupling capacitors, calculate the frequency of concern based on logic family and clock speed used. Select capacitance value based on the reactance that the capacitor presents to the circuit. A capacitor is capacitive up to its self-resonant frequency. Above self-resonance, the capacitor becomes increasingly inductive, and this minimizes RF decoupling. Table 3.1 illustrates the self-resonant frequency of axial or radial lead ceramic capacitors with standard 0.25 inch lead lengths. The self-resonant frequencies of SMT capacitors are generally a factor of 10 higher (although this beneficial multiple can be obviated by connection inductance). This is due to lower lead length inductance provided by the smaller case package size and lack of radial or axial leads.

Although Table 3.1 illustrates the self-resonant frequency of capacitors with 0.25 inch leads, keep in mind that the resonant frequency of a capacitor is also a function of lead length inductance. While the focus in this chapter is on decoupling through the use of capacitors, sometimes it is forgotten that capacitors behave like an inductor above self-resonance. Any discussion related to decoupling must consider inductance in addition to capacitance.

An inductor does not change in response like a capacitor. Instead, the magnitude of impedance changes as the frequency changes. Parasitic capacitance around an inductor can, however, cause parallel resonance and alter response. The higher the frequency of the circuit, the greater the impedance. RF current traveling through an impedance causes an RF voltage. As a result, RF current is created in the device (as related to Ohm's law: $V_{rf} = I_{rf} \times Z_{rf}$). One of the most important design concerns

Table 3.1 Self-Resonant Frequencies of Capacitors with 0.25" Leads (Assume L = 15 nH/inch)

Value	*Resonant Frequency (MHz)*
1.0 μF	2.5
0.1 μF	5.0
0.01 μF	15
0.001 μF	50
500 pF	70
100 pF	150
50 pF	230
10 pF	500

when using capacitors for decoupling lies in lead length inductance. SMT capacitors perform better at higher frequencies than radial or axial capacitors, due to less internal lead inductance in the part. Table 3.2 shows the magnitude of impedance of a 15 nH inductor versus frequency. This inductance value is caused by the lead lengths of the capacitor and the method of placement of the capacitor on a typical PCB.

Figure 3.5 shows the self-resonant frequency of various capacitor values along with different logic families [1]. It is observed that capacitors are capacitive until they approach self-resonance (null point) before going inductive. Above the point where capacitors go inductive, they proportionally cease to function for RF decoupling; however, they may still be the best source of charge for the device, even at frequencies where they are inductive. This is because the lead length of the capacitor from the internal interconnect of the capacitor plate to the mounting pad (or pin) of the PCB must be taken into consideration. Inductance is what causes capacitors to become less useful at frequencies above self-resonance for decoupling purposes.

It is also observed that certain logic families generate a greater spectrum of RF energy. This energy is generally higher in frequency than the self-resonant frequency range a decoupling capacitor presents to the circuit. For example, a 0.1 μF capacitor will usually not decouple RF currents for an "ACT or F" logic device, whereas a 0.001 μF capacitor is a more appropriate choice because of the faster edge rate (0.8–2.0 ns minimum) typical of these higher-speed components.

Effective capacitive decoupling is achieved when capacitors are properly placed on the PCB. Random placement or excessive use of capacitors is a waste of material. Sometimes, fewer capacitors strategically placed

Table 3.2 Magnitude of Impedance of a 15 nH Inductor vs. Frequency

Frequency (MHz)	*Z (Ohms)*
0.1	0.01
0.5	0.05
1.0	0.10
10.0	1.0
20.0	1.9
30.0	2.8
40.0	3.8
50.0	4.7
60.0	5.7
70.0	6.6
80.0	7.5
90.0	8.5
100.0	9.4

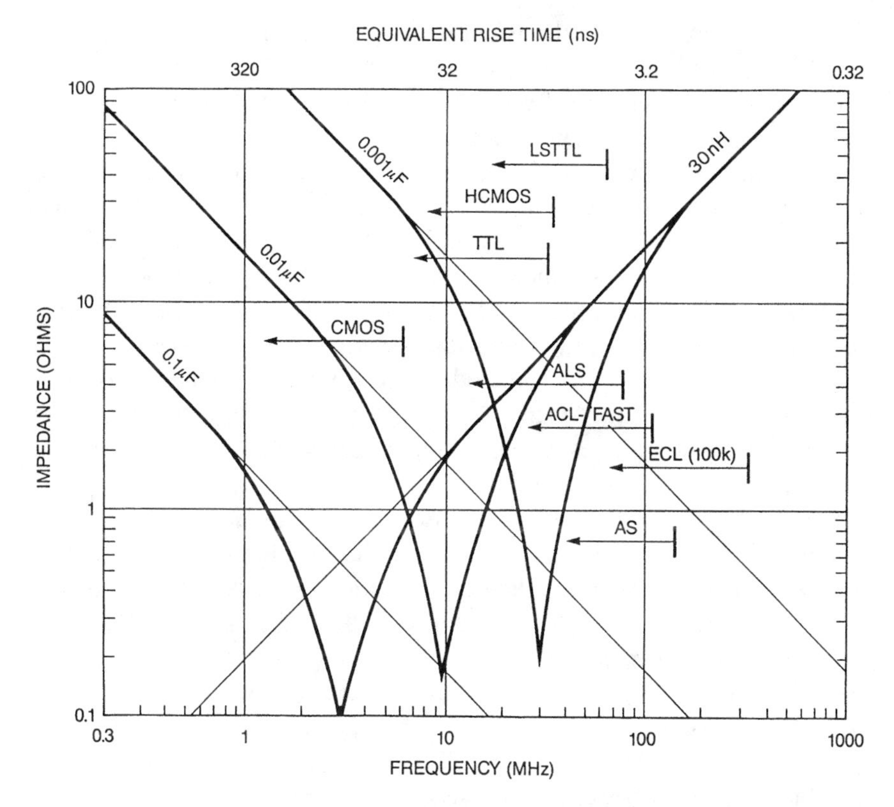

Fig. 3.5 Self-resonant frequency of capacitors vs. logic families with 1/4-inch leads. Source: *Noise Reduction Techniques in Electronic Systems,* by H. Ott. Copyright © John Wiley & Sons, reprinted with permission.

perform best for decoupling. In certain applications, two capacitors in parallel are required to provide greater spectral bandwidth of RF suppression. These parallel capacitors must differ by two orders of magnitude of value (e.g., 0.1 and 0.001 μF), or 100×, for optimal performance. Parallel capacitors are discussed in section 3.4.

A benefit of using multilayer PCBs is the placement of the power and ground planes adjacent to each other. The physical relationship of these two planes creates one large decoupling capacitor. This capacitor (zero cost in material) usually provides adequate decoupling for low-speed (slow edge rate) designs. If components have signal edges (t_r) slower than 10 ns (i.e., standard TTL logic), use of high-performance, self-resonant frequency decoupling capacitors is generally not required; however, bulk capacitors are still needed to maintain proper voltage levels for performance reasons, and values such as 0.1 μF are appropriate on the device power pins.

Another factor to consider when using power and ground planes as a primary decoupling capacitor is the self-resonant frequency of these planes. If the self-resonant frequency of the power and ground planes is the same as the self-resonant frequency of the lumped total of the decoupling capacitors placed on the board, there will be a sharp resonance where these two frequencies meet. No longer would there be a wide spectral distribution of decoupling. If a clock harmonic is at the same frequency as this sharp resonance, the board will act as if little decoupling exists at higher frequencies. When this situation occurs, the PCB may become an unintentional radiator with possible noncompliance with EMI requirements. Should this occur, additional decoupling capacitors (with a different self-resonant frequency) will be required to shift the resonance of the PCB's power planes. Use of parallel capacitors in conjunction with power planes is discussed in sections 3.4 and 3.5.

One simple method to change the self-resonant frequency of the power and ground planes is to change the spacing distance between these planes. Increasing or decreasing the height separation or relocation within the layer stackup will change the capacitance value of the assembly. Equation (3.5), presented later on, provides this calculation. One disadvantage of using this technique is that the impedance of the signal routing layers may also change. Many multilayer PCBs generally have a self-resonant frequency between 200 and 400 MHz. Use of the 20-H rule (see section 2.2) may increase the self-resonant frequency by a factor of 2 to 3.

In the past, slower logic devices fell well below the spectrum of the self-resonant frequency of the PCB's power and ground planes. Logic devices used in newer, high-technology designs approach this critical resonant frequency. When both the impedance of the power planes and the decoupling capacitors approach the same resonant frequency, severe performance deterioration occurs. This degraded high-frequency impedance will result in serious EMI problems. Basically, the assembled PCB becomes an unintentional transmitter. (The PCB is not really the transmitter; rather, the highly repetitive circuits or clocks are the sources of RF energy.) Decoupling will not solve this type of problem (due to the resonance of the decoupling), and containment measures must be employed.

3.3 CAPACITOR VALUE SELECTION

Selection of a capacitor based on past lower-frequency experience generally will not provide proper bypassing and decoupling when used with high-technology, high-speed designs. Consideration of resonance, placement on the PCB, lead length inductance, existence of power planes, and the like, must all be included when selecting a capacitor or capacitor combination.

For bulk bypass capacitors, the following procedure is recommended [2]:

1. Determine the maximum current draw anticipated on the board. Assume all gates switch at the same time. Include the effect of power surges by logic crossover (cross-conduction currents).
2. Calculate the maximum amount of power supply noise permitted by the logic devices used. Factor in a safety margin.
3. Determine the maximum common-path impedance tolerable. This is determined by

$$X_{max} = \frac{\Delta V}{\Delta I} \tag{3.2}$$

4. If solid planes are used, X_{max} is allocated to the connection between power and ground.
5. Calculate the inductance of the interconnect cabling from the power supply to the board. Add this value to X_{max} to determine the frequency below which the power supply wiring is adequate. If all gates switch simultaneously, power supply noise will be less than ΔV.

$$F_{psw} = \frac{X_{max}}{2\pi L_{psw}} \tag{3.3}$$

6. Below frequency F_{psw}, the power supply wiring is fine. Above F_{psw}, bulk capacitors are required. Calculate the value of a capacitor that has an impedance X_{max} at frequency F_{psw}.

$$C_{bypass} = \frac{1}{2\pi F_{psw} X_{max}} \tag{3.4}$$

Example 3.1 Calculation of Bulk Capacitor [2]

Assume a PCB with 100 CMOS gates (N), each switching 10 pF (C) loads in 5 ns time periods. Power supply inductance is 100 nH. Determine the correct value for the bypass capacitor.

$$\Delta I = NC\frac{\Delta V}{\Delta t} = 100\,(10\text{ pF})\frac{5\text{ V}}{5\text{ ns}} = 1.0\text{ A (worst-case peak surge)}$$

$$\Delta V = 0.100 \text{ V (from noise margin budget)}$$

$$X_{max} = \frac{\Delta V}{\Delta I} = 0.1\ \Omega$$

$$L_{psw} = 100 \text{ nH}$$

$$F_{psw} = \frac{X_{max}}{2\pi L_{psw}} = \frac{0.1\ \Omega}{2\pi 100 \text{ nH}} = 159 \text{ kHz}$$

$$C_{bulk} = \frac{1}{2\pi F_{psw} X_{max}} = 10\ \mu\text{F}$$

Capacitors commonly found on PCBs for bypassing (bulk) are in the range of 10–100 μF.

Capacitance required for decoupling power plane RF currents due to the switching energy of components can be determined from a knowledge of the resonant frequency of the logic circuits to be decoupled. Using the equations shown in Figures 3.2 through 3.4, it becomes a matter of simple arithmetic. The hardest part of calculating the value of the capacitor is knowing the inductance value of the capacitor's leads. This inductance is known as the equivalent series inductance (ESL). A discussion of ESL is found in Section 3.6. If the ESL is not known, an impedance meter or network analyzer may be used to measure the ESL value.

3.4 PARALLEL CAPACITORS

Research on the effectiveness of multiple decoupling capacitors shows that parallel decoupling may not be significantly effective, and that at high frequencies, "only" a 6 dB improvement may occur over the use of a single large-value capacitor [3]. Although 6 dB appears to be a small number for suppression of RF current, this 6 dB may be all that is required to bring a noncompliant product into compliance with international EMI specifications. According to Paul,

> *Above the self-resonant frequency of the larger value capacitor where its impedance increases with frequency (inductive), the impedance of the smaller capacitor is decreasing (capacitive). At some point, the impedance of the smaller value capacitor will be smaller than that of the larger value capacitor and will dominate, thereby giving a smaller net impedance than that of the larger value capacitor alone.*

This fact becomes evident when one examines the effects of mutual inductance and lead lengths of the decoupling capacitors. A rigorous examination of mutual inductance is found in many excellent EMC textbooks. Mutual inductance is beyond the scope of this design guide.

This 6 dB improvement is basically the result of lower lead inductance provided by the capacitors in parallel. Two capacitors in parallel will have a total capacitance value equal to the parallel value of both capacitors. However, there are now two sets of parallel leads from the internal plates of the capacitor. These two sets of leads provide greater trace width than would be available if only one set of leads were provided. With a wider trace width, there exists less lead length inductance. This reduced lead length inductance is a significant reason why parallel decoupling capacitors work as well as they do.

Figure 3.6 shows a plot of two bypass capacitors, values 0.01 μF and 100 pF, installed individually and in parallel.

> *Between the self-resonant frequency of the larger value capacitor, f_1, and the self-resonant frequency of the smaller value capacitor, f_3, the impedance of the larger value capacitor is essentially inductive, whereas the impedance of the smaller value capacitor is capacitive. In this frequency range there exists a parallel resonant LC circuit and we should therefore expect to find an infinite impedance of the parallel combination. Around this resonant point, the impedance of the parallel combination is actually larger than the impedance of either isolated capacitor! [3]*

To remove RF current generated by components switching all signal pins simultaneously, it is common practice to place two capacitors in parallel (i.e., 0.1 μF and 0.001 μF) immediately adjacent to each power pin. Capacitance values must differ by two orders of magnitude, or 100×. The total capacitance of parallel capacitors is not important. Parallel reactance provided by the parallel capacitors (due to self-resonant frequency) is the important item. (Refer to Tables 3.1 and 3.2.)

To optimize the effects of parallel bypassing and to allow use of only "one value of capacitor," reduction in capacitor lead length inductance is required. This is practically impossible, since a finite length of lead inductance will always exist when installing the capacitor on the PCB. Note that the "lead length" also must include the length of the vias connecting the capacitors to the planes. The shorter lead length provided, from single or parallel decoupling, the greater the performance of the decoupling capacitor.

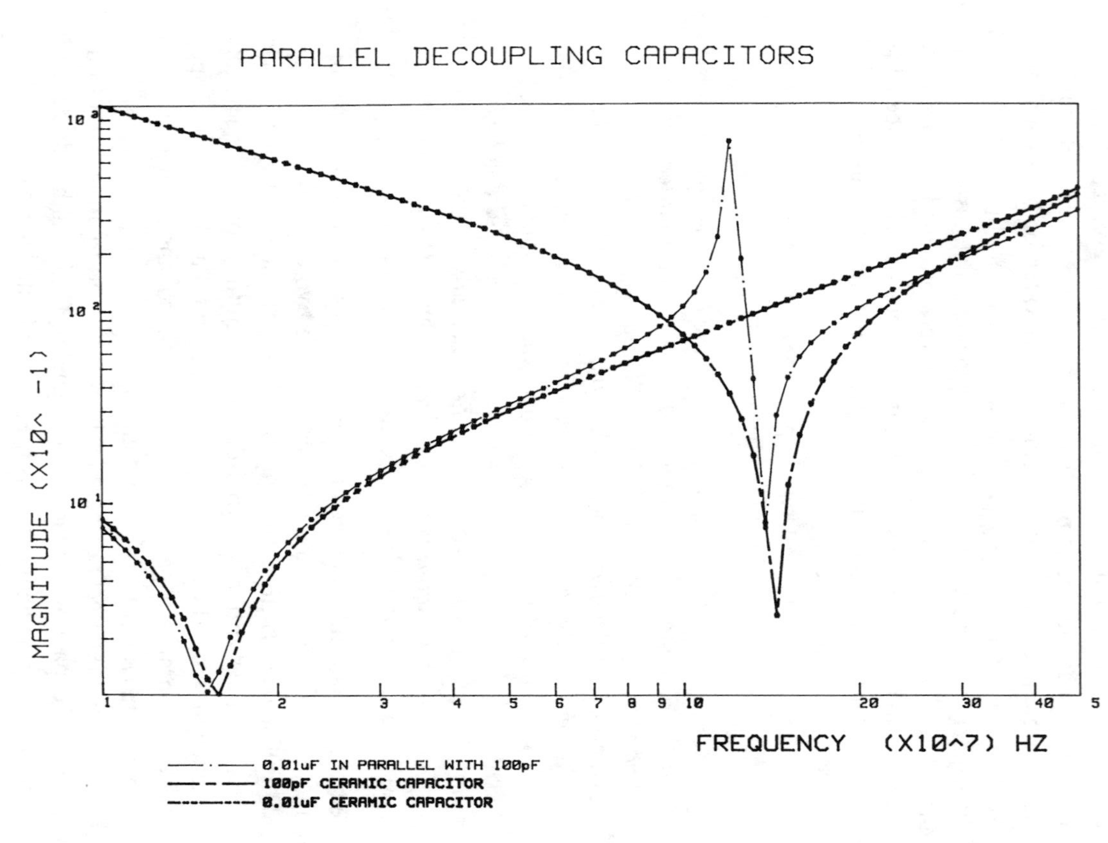

Fig. 3.6 Resonance of parallel capacitors. Source: reprinted from Ref. [3] with permission

3.5 POWER AND GROUND PLANE CAPACITANCE

The effects of the internal power and ground planes inside the PCB are not considered in Fig. 3.6. However, this triple-bypassing effect is illustrated in Fig. 3.7. Power and ground planes have very little "lead" inductance equivalence and no equivalent series resistance (ESR), as discussed more fully in Section 3.6. Use of these power planes as a decoupling capacitor reduces RF energy at frequencies generally in the higher ranges.

Capacitance will always exist between the voltage and ground planes. Depending on the thickness of the core material, dielectric constant of the material, and placement of the power planes in the board stackup, various values of internal capacitance can exist. Network analysis, mathematical calculations, or modeling will reveal actual capacitance of the power planes. This is in addition to determining the impedance of all circuit planes

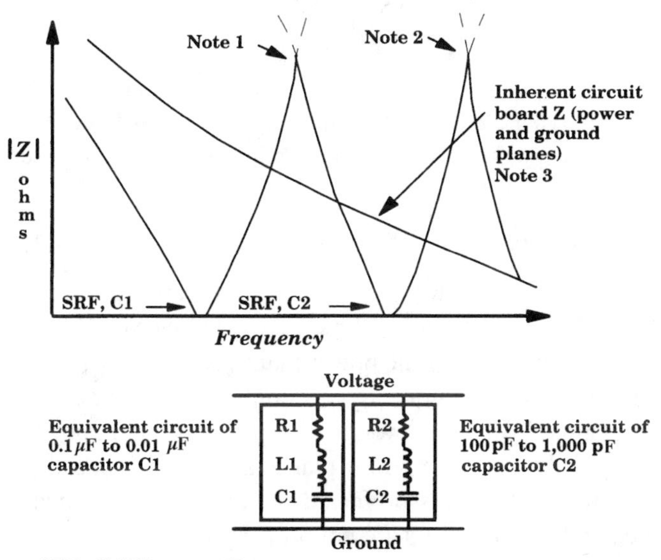

SRF = Self-Resonant Frequency

Note 1 - Sharp overshoot Z, caused by the parallel interaction of L1 within C1 with capacitor C2, and similar in interaction of L2 within C2 to C1 (infinite impedance).

Note 2 - Sharp overshoot Z, caused by the parallel interaction of L2 within C2 with inherent impedance of power plane structure (infinite impedance).

Note 3 - Decoupling shape for power and ground plane will be different from that shown due to the distance separation between the planes and their self-resonant frequency (board dependent).

Fig. 3.7 Parallel decoupling effects combined with power plane capacitance (Note: drawing for poles and zeroes (resonant values) not to scale)

and the self-resonant frequency of the total assembly as potential RF radiators. This value of capacitance is easily calculated by Eqs. (3.5) and (3.6).

$$C = \frac{\varepsilon_o \varepsilon_r A}{d} = \frac{\varepsilon A}{d} \tag{3.5}$$

where

ε = permittivity of the medium between capacitor plates, in F/m

A = area of the power planes, in m^2

d = separation of the plates, in m

C = capacitance between the power planes, in pF

Introducing relative dielectric constant ε_r and the value of ε_o, we obtain the capacitance for the parallel-plate capacitor; namely, the power and ground plane combination.

$$C = 8.85 \frac{A\varepsilon_r}{d} \text{ (pF)} \tag{3.6}$$

where

ε_r = the relative dielectric constant of the medium between the plates, typically ≈ 4.5 (varies for linear material, usually between 1 and 10)

ε_o = dielectric constant of free space, $(1/36\pi) \times 10^{-9} = 8.84 \times 10^{-12}$

Equations (3.5) and (3.6) show that the power and ground planes, separated by 0.01 inches of FR-4 material, will have a capacitance of 100 pF/in^2.

Because of the efficiency of the power planes as a decoupling capacitor, use of high self-resonant frequency decoupling capacitors may not be required for standard TTL or low-speed logic. This optimum efficiency exists, however, only when the power planes are closely spaced (less than 0.01 inches with 0.005 inches preferred for high-speed applications). If additional decoupling capacitors are not properly chosen, the power planes may go inductive below the lower cut-in range of the higher self-resonant frequency decoupling capacitor. With this gap in resonance, a pole is generated, causing undesirable effects on RF suppression. At this point, RF suppression techniques on the PCB become ineffective, and containment measures must be used—at a much greater expense.

3.6 CAPACITOR LEAD LENGTH INDUCTANCE

All capacitors have lead length inductance, and all vias add to this inductance. Lead inductance must be minimized at all times. When a signal trace plus lead length inductance is combined, a higher impedance mismatch will exist between the component's ground and the ground planes. With this impedance mismatch, a voltage gradient is created between these two sources, thus creating RF fields. RF fields cause RF emissions on PCBs; hence, decoupling capacitors must be designed for minimum inductive lead length, including via and escape trace lengths.

In a capacitor, the dielectric material also determines the self-resonant frequency of operation. The dielectric material is very temperature sensitive. The capacitance value of the capacitor will change in relation to the ambient temperature provided to its case. At certain temperatures, the capacitance may change substantially and may result in improper performance, or no performance at all when used as a bypass or decoupling element. The more stable the dielectric material, the better performance of the capacitor.

In addition to the sensitivity of the dielectric material to temperature, the equivalent series inductance (ESL) and the equivalent series resistance (ESR) must be low at the desired frequency of operation. ESL behaves like a parasitic inductor, while ESR acts like a parasitic resistor, both in series with the capacitor. ESL is not a factor in today's small SMT capacitors. Radial and axial lead devices have a large ESL value. Together, ESL and ESR degrade a capacitor's effectiveness as a bypass element. When selecting a capacitor, choose a capacitor vendor that publishes actual ESL and ESR values in their data sheets. Random selection of a standard capacitor may result in improper performance if ESL and ESR are too high. Most vendors of capacitors do not publish ESL and ESR values, so beware of this selection parameter when choosing capacitors for use in high-speed, high-technology PCBs.

Because surface mount capacitors have small values of ESL and ESR, their use is preferred over radial or axial types. Typically, ESL should be < 10 nH, and ESR should be 0.5 Ω or less. For decoupling capacitors, capacitance tolerance is not as important as the temperature stability, dielectric constant, ESL, ESR, and self-resonant frequency [4].

3.7 PLACEMENT

3.7.1 Power planes

Multilayer PCBs generally contain one or more pairs of voltage and ground planes. Power planes function as a low-inductance capacitor that contains RF currents generated from components and traces. Multiple chassis ground-stitch connections to all ground planes minimize voltage gradients between board and chassis. These gradients also are a major source of common-mode RF fields. This is in addition to sourcing RF currents to chassis ground. In many cases, multiple ground connections are not possible, especially in card cage designs. In such situations, consideration must be taken to analyze and determine where RF loops will occur and how to optimize grounding of the power planes.

Power planes provide for power flux cancellation, in addition to decoupling RF currents created from power fluctuations of components and noise injected into the planes from the power supply. An image plane is a solid copper plane at voltage or ground potential, located adjacent to a signal routing plane. RF currents generated by traces on the signal plane will mirror image themselves in this adjacent solid metal plane. To remove common-mode RF currents in the PCB, all routing (signal) layers must be physically adjacent to an image plane. Refer to Chapter 2 for a detailed discussion of image planes.

3.7.2 Capacitors

Before determining where to locate the capacitors, the physical structure of the PCB must be understood. Figure 3.8 shows the electrical equivalent circuit of a PCB. In this figure, observe the loops that exist between power and ground caused by traces, IC wire bonds, lead frames of components, socket pins, component interconnect leads, and decoupling capacitor. The key to effective decoupling is to minimize R_2, L_2, R'_2, L'_2, R_3, L_3, R'_3, L'_3 and R_4, L_4, R'_4 and L'_4. Placement of power and ground pins in the center of the component assist in reducing R_4, L_4, R'_4 and L'_4.

In Fig. 3.8, it is evident that EMI is a function of loop geometry and frequency; hence, the smallest closed loop area is desired. We acquire this small area by placing a local decoupling capacitor for current storage adjacent to the power pins of the IC. It is mandatory that the decoupling loop impedance be much lower than the rest of the power distribution system. This low impedance will cause the high-frequency components in the traces and circuit components to remain almost entirely within this closed loop. As a result, lower EMI emissions are observed.

If the impedance of the loop is larger than the rest of the system, some fraction of the high-frequency RF component will transfer to the larger loop formed by the power distribution system. With this situation, RF currents are generated and, hence, EMI emissions. This situation is illustrated in Fig. 3.9.

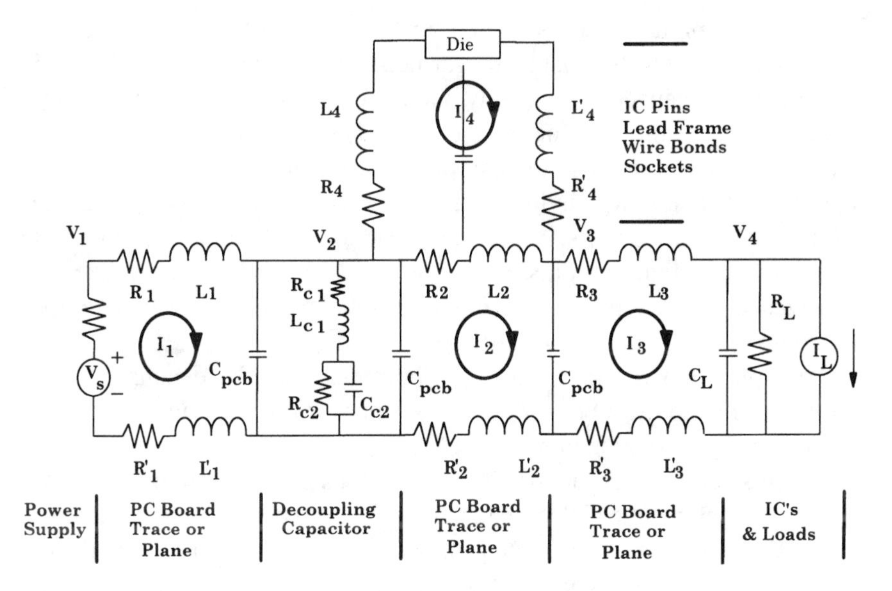

Fig. 3.8 Equivalent circuit of a PCB

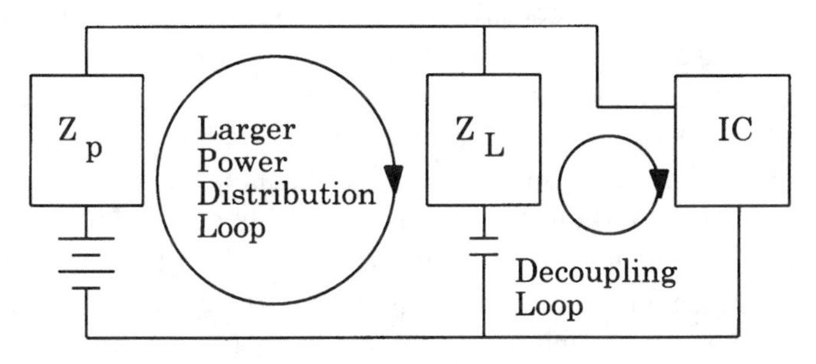

Fig. 3.9 Power distribution model for loop control

To summarize,

The important parameter when using decoupling capacitors is to minimize lead length inductance and locate as close as possible to the component.

Decoupling capacitors must be provided for every component with edges (t_r) faster than 5 ns and should be provided, placement wise, for "every component." Making provisions for decoupling capacitors is a necessity because future EMI testing may indicate a requirement for these capacitors. During testing, it may be possible to remove some or all of these excess capacitors. Having to add capacitors to an assembled board is difficult, if not impossible. Today, CMOS, ECL, and other fast logic families require additional decoupling capacitors.

If a decoupling capacitor must be provided to a through-hole device after assembly, retrofit must be performed. Several manufacturers provide a decoupling capacitor assembly using a flat, level construction that resides between the component and PCB. This flat pack shares the same power and ground pins of the components. Because these capacitors are flat in construction, R_{c1} and L_{c1} are reduced (refer to Fig. 3.8) in comparison to leaded capacitors. Since the capacitor and component share the same power and ground pins, R_3, L_3, R'_3 and L'_3 also are reduced. Some impedance will remain which cannot be removed. The most widely used board level retrofit capacitors are known as Micro-Q™.*

Poor planning in anticipation for possible use of a decoupling may require use of Micro-Q retrofit devices. There is no equivalent retrofit for SMT packaged components.

When selecting a capacitor, one should consider not only the self-resonant frequency but the dielectric material as well. The most commonly used material is Z5U (barium titanate ceramic). This material has a high dielectric constant. This constant allows small capacitors to have large capacitance values with self-resonant frequencies from 1 MHz to 20 MHz, depending on design and construction. Above self-resonance, performance of Z5U decreases as the loss factor of the dielectric becomes dominant, which limits its usefulness to approximately 50 MHz.

Another dielectric material commonly used is NPO (strontium titanate). This material has a much better high-frequency performance due to its low dielectric constant. Capacitors using this material are unsuitable for decoupling below 10 MHz. NPO is also a more temperature stable dielectric. Capacitance value (and self-resonant frequency) is less likely to change when the capacitor is subjected to changes in ambient temperature.

A problem observed when Z5U and NPO are provided in parallel is that the higher dielectric material, Z5U, can damp the resonance of the more frequency-stable, low-dielectric constant material, NPO. For EMI problems below 50 MHz, it is better to use only a good, low inductance Z5U (or equivalent) capacitor. This is because Z5U combines excellent low-frequency decoupling with reduction in radiated emissions.

Placement of 1 nF (1000 pF) capacitors (capacitors with a very high self-resonant frequency) on a one-inch center grid throughout the PCB also provides additional protection from reflections and RF currents generated by both signal traces and power planes, especially if a high-density PCB stackup is used [4]. It is not the exact location that counts in placement of these additional decoupling capacitors. A lumped model

* Micro-Q is a trademark of Circuit Components, Inc. (formerly Rogers Corporation).

analysis of the PCB will show that the capacitors will still function as needed, regardless of where the devices are actually placed for overall decoupling performance. Depending on the resonant structure of the board, values of the capacitors placed in the grid may be as small as 30 to 40 pF.

VLSI and high-speed components (e.g., F, ACT, BCT, CMOS, and ECL logic families) may require decoupling capacitors in parallel. As edge rates of components become steeper, a greater spectral distribution of RF currents is created. Parallel capacitors generally provide optimal bypassing of power plane noise in addition to containing high-frequency RF energy. Place multiple paired sets of capacitors between the power and ground pins of VLSI components, located around all four sides. These high-frequency decoupling capacitors are typically rated 0.1 μF in parallel with 0.001 μF. For 50 MHz systems and higher clock frequencies, use a parallel combination of 0.01 μF and 100 pF components.

While the focus in this chapter is on multilayer boards, single- and double-sided boards also require decoupling. Figure 3.10 illustrates correct and incorrect ways of locating decoupling capacitors for a two-layer assembly. When placing decoupling capacitors, ground loop control must be considered at all times. When using multilayer boards with internal power and ground planes, placement of the decoupling capacitor may be anywhere in the vicinity of the component's power pins. This is due to the lumped distributed capacitance of the power planes and the fact that the components themselves must via down to the power and ground plane—the same as the decoupling capacitor.

In Fig. 3.10, V_{gnd} is LdI/dt induced noise in the ground trace flowing in the decoupling capacitor. This V_{gnd} now drives the ground structure of the board and contributes to the overall common-mode voltage across the entire board. Thus one should minimize the ground path that is common with the board's ground structure and the decoupling capacitor.

3.7.3 Bulk capacitors

Bulk capacitors provide dc voltage and current to components when the devices are switching all data, address, and control signals simultaneously under maximum capacitive load. Components have a tendency to cause current fluctuations on the power planes. These fluctuations can cause improper performance of components due to voltage sags. Bulk capacitors provide energy storage for circuits to maintain voltage and current requirements. Bulk capacitors play no significant role in EMI control.

Optimal placement for components with the power pins in the middle of the device.

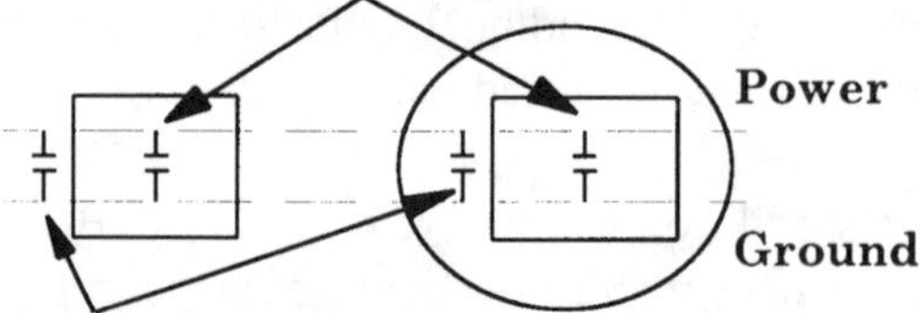

Optimal placement for components with the power pins in the opposite corners

Recommended Power Rail Layout
Best Implementation Technique
for single and double sided
printed circuit boards

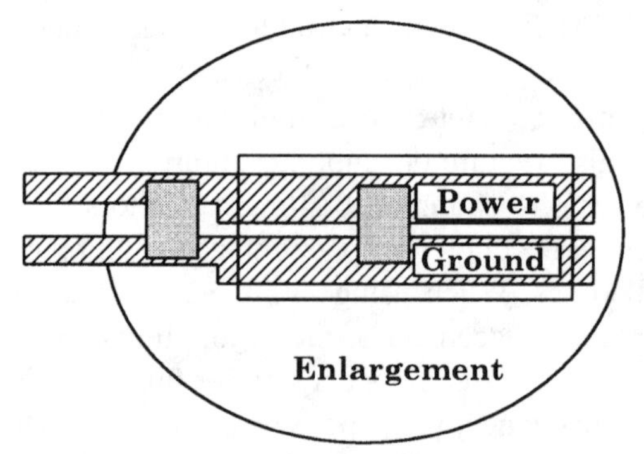

Area under component should be filled in completely to avoid coupling around the capacitor

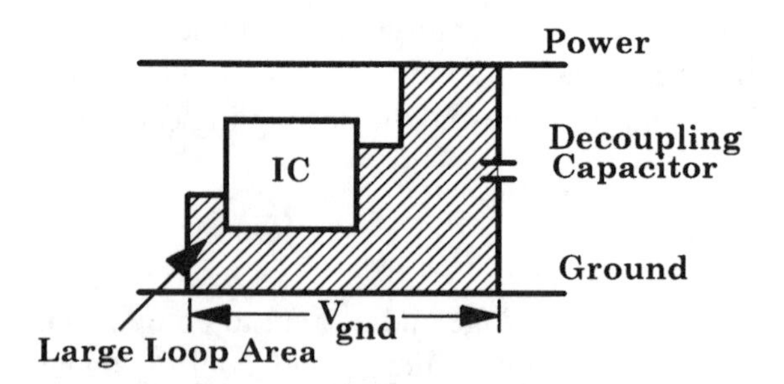

Commonly Used Power Rail Layout
Poor Implementation Technique
for single and double sided
printed circuit boards.

Fig. 3.10 Decoupling capacitor placement, two-layer PCB

Bulk capacitors (tantalum dielectric) should be used in addition to high self-resonant frequency decoupling capacitors to provide dc power for components and power plane RF modulation. Place one bulk capacitor for every two LSI and VLSI components, in addition to the following locations:

- power entry connector from the power supply to the printed circuit board
- power terminals on I/O connectors for daughter cards, peripheral devices, and secondary circuits
- adjacent to power-consuming circuits and components
- the farthest location from the input power connectors
- high-density component placement remote from the dc input power connector
- adjacent to clock generation circuits

When using bulk capacitors, calculate the voltage rating such that the nominal voltage equals 50 percent of the capacitor's voltage rating to prevent the capacitor from self-destruction if a voltage surge occurs. For example, with power at 5 V, use a capacitor with a minimum 10 V rating.

Table 3.3 shows the typical number of capacitors required for some popular logic families. This chart is based on the maximum allowable power drop, which is equal to 25 percent of the noise immunity level of the circuit being decoupled. Note that for standard CMOS logic, this table is conservative, since the trace wiring to the components cannot provide the required peak surge current without excessive voltage drop.

Table 3.3 Number of Decoupling Capacitors for Selected Logic Families

	Peak Transient Current Requirement (mA)		
Logic Family	*Gate Overcurrent (mA)*	*1 Gate Drive (mA)*	*No. of Decoupling Capacitors for Fan-Out of 5 Gates + 10 cm Trace Length*
CMOS	1	0.3	0.6
TTL	16	1.7	2.6
LS-TTL	8	2.5	2.0
HCMOS	15	5.5	1.2
STTL	30	5.0	1.8
FAST	15	5.5	1.8
ECL	1	1.2	0.9

Source: reprinted from Ref. [5] by permission.

Memory arrays require additional bulk capacitors due to the extra current required for proper operation during a refresh cycle. The same is true for VLSI components with high pin counts. High-density pin grid array (PGA) modules also must have additional bulk capacitors provided, especially when all signal, address, and control pins switch simultaneously under maximum capacitive load.

Using Eq. (3.1) to calculate the peak surge current consumed by many capacitors, it is observed that more is not necessarily better. An excessive number of capacitors can draw a large amount of current and thereby place a strain on the power supply.

3.8 REFERENCES

1. Ott, H. 1988. *Noise Reduction Techniques in Electronic Systems,* 2nd ed. New York: John Wiley & Sons.
2. Johnson, H., and M. Graham. 1993. Reprinted by permission from *High Speed Digital Design.* Englewood Cliffs, N.J.: PTR Prentice Hall.
3. Paul, C.R. 1992. Effectiveness of multiple decoupling capacitors. *IEEE Transactions on Electromagnetic Compatibility* EMC-34:130–133.
4. Montrose, M.I. 1991. Overview on design techniques for PCB layout used in high technology products. *Proceedings of the IEEE International Symposium on Electromagnetic Compatibility,* 61–66.
5. Mardiguian, M. 1992. *Controlling Radiated Emissions by Design.* New York: Van Nostrand Reinhold.

4

Clock Circuits

To a significant degree, clock generators, associated components, and distribution lines account for the emissions generated on a PCB. A clock circuit area is defined as the functional area that physically contains the clock oscillator and/or its buffers, drivers, and associated components, both active and passive.

The techniques outlined in this chapter are appropriate when designing clock circuits. RF emissions are directly related to the rise and fall times (edges) of active components. To determine the highest RF generated frequency, Eq. (4.1) is used. This equation does not take into account harmonics that are created from the primary frequency.

$$f_{max} = \frac{1}{\pi \times t_r} \tag{4.1}$$

where

f_{max} = maximum generated RF frequency

t_r = pulse (or edge) rise time

For example, a 2 ns edge rate, typical of common clock drivers and components, can be expected to radiate significant RF energy up to 160 MHz, which falls off rapidly above that level. The possible significant RF spectrum is $10 \times f_{max}$, or 1.6 GHz, which includes the harmonic content of the main frequency component.

Clock traces always must be *manually* routed before automatic routing of the remaining traces—no exceptions. Upon successful manual routing of these clock traces, the rest of the PCB can be routed automatically.

4.1 PLACEMENT

Locate clock circuits near the center and/or a ground stitch location (to chassis ground) on the PCB rather than along the perimeter or near the I/O section. If the clock goes off the board to a daughter card, ribbon cable, or other peripheral, locate the clock circuit distant from this interconnect, with the clock trace terminated directly at the connector. It is imperative that this be made up of a point-to-point radial. Termination of clock lines at interconnect locations enhances signal quality by providing proper termination instead of an unterminated clock line that is open ended and acts as a monopole antenna. In addition to termination of this clock trace, suppression of RF currents coupling into other areas susceptible to RF corruption, discussed later in this chapter, is enhanced. Install oscillators and crystals directly on the PCB, and

Do not use sockets!

Sockets add lead length inductance (LdI/dt) and provide an additional path for RF currents and harmonics to radiate or couple into areas both internal to the product and outside environment. This increases common-mode EMI levels due to ground bounce across the inductance.

Place only traces associated with the clock circuitry in the clock generation area. Avoid placing any other traces "near, under, or through" this clock circuit on an adjacent signal routing layer. This route keep-out area is required only for a second microstrip signal layer between the top (or component) layer and the first image plane (power or ground). If a trace must be routed on a two- or four-layer board through this area, route this trace only on the solder (bottom) side. Do not route traces in the vicinity of the oscillator output pin or directly under the oscillator!

Allow for possible use of a Faraday cage (a metal can that encompasses devices 100 percent) to be installed around the entire clock circuit area with a ground trace circumscribing the zone (except for the point of signal trace exit, if required). This shield must be of an RF type similar to those used in UHF and microwave applications. This is best accomplished by placing ground vias in the board around the device. Always provide for additional source of grounding the metal case of oscillator modules, as the ground pin of the oscillator is usually not sufficient to source RF currents created internal to the package to ground. This ground pin is provided for dc voltage reference and was not designed as a low impedance path to source RF currents to ground that are acceptable for a shield.

Use oscillators for frequencies above 5 MHz or clock edges faster than 5 ns instead of discrete components and crystals. Exceptions do exist for specialized circuits.

4.2 LOCALIZED GROUND PLANES

Locate oscillators, crystals, and all clock support circuitry (i.e., buffers, drivers, etc.) over a single localized ground plane. This localized ground plane is on the component (top) layer of the PCB and ties directly into the main ground plane(s) of the PCB through both the oscillator ground pin and a minimum of two additional vias. This ground plane should also be positioned next to and connected to a ground stitch location. Do not place solder mask on this localized ground plane. Solder mask changes the dielectric constant between the top layer and localized ground plane and could minimize desired RF coupling between source and ground. An example of this localized ground plane is shown in Fig. 4.1.

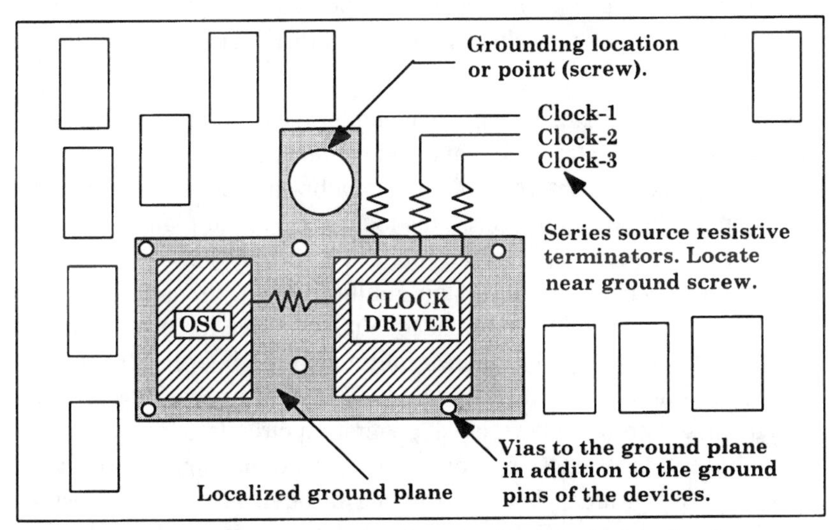

Note 1: Do not run any traces on layer 1 through the localized ground plane.

Note 2: If two microstrip layers exist, do not route any traces on layer two of the localized ground plane (route keep-out area).

Note 3: The localized ground plane is solid copper plane without solder mask bonded to the main ground plane(s) by vias "and" a mask bonded to a ground stitch (screw or equivalent).

Fig. 4.1 Localized ground plane

The following are the main reasons for placing a localized ground plane under the clock area:

- Circuitry inside the oscillator creates RF currents. If the oscillator package is a metal can, the dc power pin is relied upon for both dc voltage reference and a path for RF currents to be sourced (or sunk) to ground from the oscillator circuitry. Depending on the type of oscillator chosen, (CMOS, TTL, ECL, etc.), the RF currents created internal to the package can become so excessive that the ground pin is unable to source this large LdI/dt current (L from the pin lead) to ground. As a result, the metal case becomes a monotropic antenna. The nearest image or ground plane (internal to the PCB) is sometimes two or more layers away and is thus inefficient as a radiated coupling path for RF currents to ground.
- If the oscillator is a surface-mount device, the situation mentioned above is made worse because SMT packages are usually plastic. RF currents created internal to the package can radiate into free space and couple to other components. The high impedance of the PCB material, relative to the impedance of the ground pin of the oscillator, prevents RF currents from being sourced to ground. SMT packages will always radiate more RF energy than a metalized case.
- Placing a localized ground plane under the oscillator and clock circuits provides an image plane that captures common-mode RF currents generated internal to the oscillator and related circuitry, thus minimizing RF emissions. This localized ground plane is also "RF hot." To contain differential-mode RF current that is also sourced to the localized ground plane, multiple connections to all system ground planes must be provided. Vias from the localized plane, on layer 1, to the ground planes internal to the board will provide a lower impedance path to ground. To enhance performance of this localized ground plane, clock generation circuits should be located adjacent to a chassis ground (stitch) connection. Connect this localized ground plane to a plated through hole, 360°. Ensure a low-impedance RF bonding connection to ground. Connection through traces to a ground location can defeat a low-impedance ground connection, and while thermal relief "wagon wheel" connections usually are acceptable, they also degrade the performance of the connection.
- When using a localized ground plane, *do not run traces through this plane*! This violates the functionality of an image plane. If a

trace travels through a localized ground plane, the potential for small ground loops or discontinuities exists. These ground loops can generate problems in the higher frequency range. Why install a plane when you defeat its functional use by running traces through it?

- Support logic circuitry (clock drivers, buffers, etc.) must be located adjacent to the oscillator. Extend this localized ground plane to include this support circuitry. Generally, an oscillator drives a clock buffer. This buffer is usually a super-high-speed, fast-edge-rate device. Because of the functional characteristics of this driver, RF currents will be created at harmonics of the primary clock frequency. With a large voltage swing and drive current injected onto the signal trace, both common-mode and differential-mode RF currents will exist. These currents can cause functionality problems and possible noncompliance to EMC requirements.

4.3 IMPEDANCE CONTROL

Clock traces should be impedance controlled. Calculate for proper trace width and distance separation to the nearest plane. Calculation of trace impedance for microstrip and stripline implementation is shown in Eqs. (4.2) through (4.8). Board manufacturers and CAD programs can easily perform these calculations for you. If necessary, consult board manufacturers for assistance in designing your PCB to determine trace width and distance spacing between planes for optimal performance.

For microstrip topology:

The approximate formula for surface microstrip impedance is:

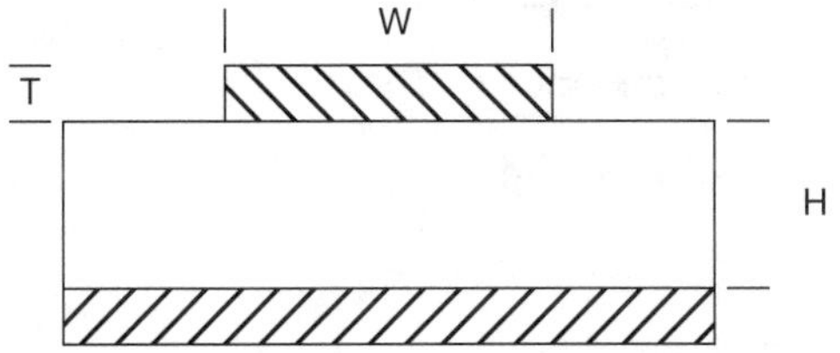

W

T

H

$$Z_o = \left(\frac{87}{\sqrt{E_r + 1.414}}\right)\ln\left(\frac{5.98H}{0.8W + T}\right) \tag{4.2}$$

The approximate formula for embedded microstrip impedance is:

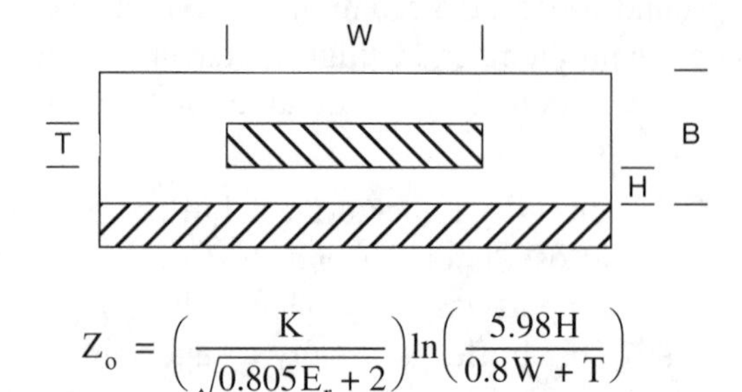

$$Z_o = \left(\frac{K}{\sqrt{0.805E_r + 2}}\right)\ln\left(\frac{5.98H}{0.8W + T}\right)$$

where $60 \le K \le 65$, or

$$Z_o = \left(\frac{87}{\sqrt{E_r' + 1.41}}\right)\ln\left(\frac{5.98H}{0.8W + T}\right) \tag{4.3}$$

where $E_r' = E_r\left[1 - e\left(\frac{-1.55B}{H}\right)\right]$

The propagation delay is:

$$t_{pd} = 1.017\sqrt{0.475E_r + 0.67} \quad \text{(ns/ft)} \tag{4.4}$$

The inductance per foot is:

$$L_o = Z_o^2 C_o \tag{4.5}$$

where

Z_o = characteristic impedance (ohms)

t_{pd} = propagation delay (ns/ft)

W = width of the trace (inches)

T = thickness of the trace (inches)

H = distance between signal trace and reference plane (inches)

B = overall dielectric thickness

E_r = dielectric constant of the planar material

C_o = Capacitance per foot

L_o = Inductance of trace

For stripline topology:

The approximate formula for single stripline impedance is:

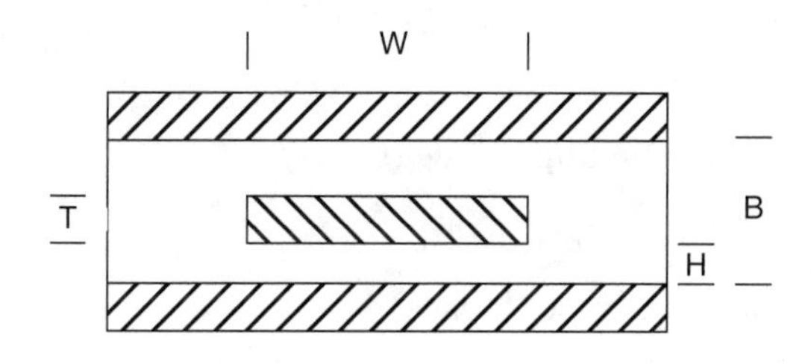

$$Z_o = \left(\frac{60}{\sqrt{E_r}}\right)\ln\left(\frac{4B}{0.67\pi W\left(0.8 + \frac{T}{W}\right)}\right) \tag{4.6}$$

The approximate formula for dual stripline impedance is:

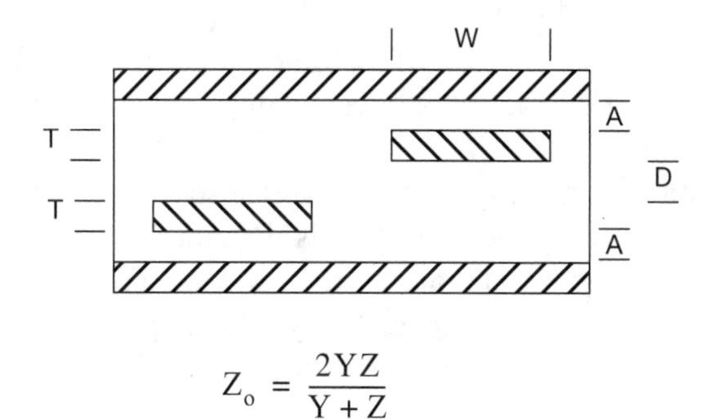

$$Z_o = \frac{2YZ}{Y + Z}$$

where

$$Y = \left(\frac{60}{\sqrt{E_r}}\right)\ln\left(\frac{8A}{0.67\pi W\,(0.8 + T/W)}\right)$$

and

$$Z = \left(\frac{60}{\sqrt{E_r}}\right)\ln\left(\frac{8\,(A + D)}{0.67\pi W\,(0.8 + T/W)}\right)$$

or:

$$Z_o = \frac{80\left[1 - \left(\frac{A}{4\,(A + D + T)}\right)\right]}{\sqrt{E_r}}\ln\left[\frac{1.9\,(2A + T)}{0.8W + T}\right] \tag{4.7}$$

The propagation delay is:

$$t_{pd} = 1.017\sqrt{E_r} \tag{4.8}$$

where

Z_o = characteristic impedance (ohms)

t_{pd} = propagation delay (ns/ft)

W = width of the trace (inches)

T = thickness of the trace (inches)

H = distance between signal trace and reference plane (inches)

B = overall dielectric thickness

A = dielectric thickness between trace and power/ground plane

E_r = dielectric constant of the planar material

Table 4.1 shows the impedance of a 10-inch × 10-inch copper metal plane internal to a PCB. Table 4.2 illustrates the impedance of a typical trace, 35 μm in height.

Maintain clock line impedance at a constant value throughout the route. Depending on application, this is generally 55 to 75 Ω ± 10 percent. If a large number of vias are used (at approximately 1 to 3 nH inductance each), these will have a tendency to change the impedance of a trace, thus causing potential EMI and functionality problems.

Table 4.1 Impedance of a 10 × 10 Inch Copper Metal Plane

Frequency (MHz)	*Skin Depth (cm)*	*Impedance (Ω/sq)*
1	6.6×10^{-3}	0.00026
10	2.1×10^{-3}	0.00082
100	6.6×10^{-4}	0.00260
1000	2.1×10^{-4}	0.00820

4.4 PROPAGATION DELAY

Propagation delay is a function of capacitance per unit length of line. This capacitance is a function of the dielectric constant, the line width, and the thickness of the dielectric between the trace and image plane. For G-10 fiberglass epoxy boards ($E_r \cong 5.0$) the propagation delay of the microstrip line is calculated to be 1.77 ns/ft. For FR-4 material with $E_r \cong 4.6$, propa-

Table 4.2 Impedance Values of Typical PCB Traces
(Source: reprinted from Ref. 1 by permission)

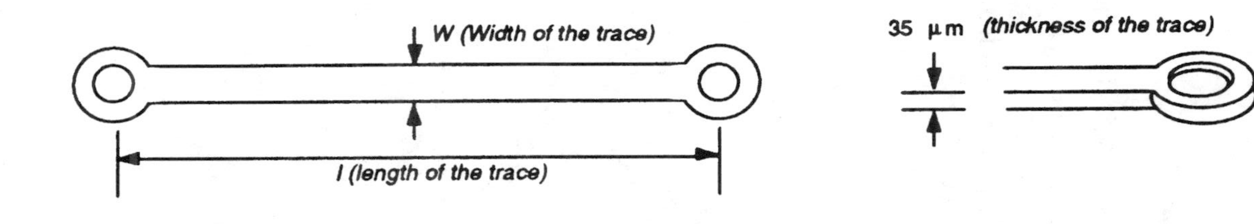

	Impedance						
	W = 1 mm				W = 3 mm		
	l = 1 cm	l = 3 cm	l = 10 cm	l = 30 cm	l = 3 cm	l = 10 cm	l = 30 cm
DC, 50 Hz to 1 kHz	5. 7 mΩ	17 mΩ	57 mΩ	170 mΩ	5.7 mΩ	19 mΩ	57 mΩ
10 kHz	5. 75 mΩ	17. 3 mΩ	58 mΩ	175 mΩ	5. 9 mΩ	20 mΩ	61 mΩ
100 kHz	7. 2 mΩ	24 mΩ	92 mΩ	310 mΩ	14 mΩ	62 mΩ	225 mΩ
300 kHz	14. 3 mΩ	54 mΩ	225 mΩ	800 mΩ	40 mΩ	175 mΩ	660 mΩ
1 MHz	44 mΩ	173 mΩ	730 mΩ	2. 6 Ω	0. 13 Ω	0. 59 Ω	2. 2 Ω
3 MHz	0. 13 Ω	0. 52 Ω	2. 17 Ω	7. 8 Ω	0. 39 Ω	1. 75 Ω	6. 5 Ω
10 MHz	0. 44 Ω	1. 7 Ω	7. 3 Ω	26 Ω	1. 3 Ω	5. 9 Ω	22 Ω
30 MHz	1. 3 Ω	5. 2 Ω	21. 7 Ω	78 Ω	3. 9 Ω	17. 5 Ω	65 Ω
100 MHz	4. 4 Ω	17 Ω	73 Ω	260 Ω	13 Ω	59 Ω	220 Ω
300 MHz	13 Ω	52 Ω	217 Ω		39 Ω	175 Ω	
1 GHz	44 Ω	170 Ω			130 Ω		

gation delay for a fully embedded microstrip is 1.72 ns/ft (0.143 ns/inch or 0.36 ns/cm). With stripline, propagation delay is 2.26 ns/ft (0.188 ns/inch or 0.0.48 ns/cm). For comparison, coaxial propagation delay is 1.52 ns/ft (0.127 ns/inch or 0.32 ns/cm). Calculation of propagation delay is performed using Eqs. (4.4) and (4.8).

4.5 CAPACITIVE LOADING

Capacitive input loading affects trace impedance and will increase with gate loading (addition of input loads). The unloaded propagation delay for a transmission line is defined by $t_{pd} = \sqrt{L_o C_o}$. If a lumped load, C_d, is placed along the line (includes all loads with their capacitance added together), the propagation delay of the signal trace will increase by a factor described by Eq. (4.9)[2]:

$$t_{pd}' = t_{pd}\sqrt{1 + \frac{C_d}{C_o}} \tag{4.9}$$

For example, assume a load of five CMOS devices are on a signal route, each with 10 pF capacitance (total of 50 pF). With a capacitance value of a 50 Ω stripline line on a glass epoxy board, with 25 mil traces and a characteristic board impedance $Z_o = 50\ \Omega$ [$t_r = 1.77$ ns/ft per Eq. (4.8)], there exists a value of $C_d = 35$. The modified propagation delay is:

$$t'_{pd} = 1.77 \text{ ns/ft}\sqrt{1 + \frac{50}{35}} = 2.75 \text{ ns/ft} \tag{4.10}$$

This says that the signal arrives at its destination 2.75 ns/ft later than expected.

The characteristic impedance of a transmission line, altered by gate loading, Z'_o is:

$$Z'_o = \frac{Z_o}{\sqrt{1 + \frac{C_d}{C_o}}} \tag{4.11}$$

where

Z_o = original line impedance (ohms)

Z'_o = modified line impedance (ohms)

C_d = input gate capacitance

C_o = characteristic capacitance of the line

For the example above

$$Z'_o = \frac{50}{\sqrt{1 + \frac{50}{35}}} = 32\ \Omega$$

Typical values of C_d are 5 pF for each ECL input, 10 pF for each CMOS input, and 10–15 pF for TTL. Typical C_o values are 2–2.5 pF/inch. Sockets and vias also add to the distributed capacitance (sockets ≈ 2 pF and vias ≈ 0.3–0.8 pF each). Given that

$$t_{pd} = \sqrt{L_o C_o} \text{ and } Z_o = \sqrt{\frac{L_o}{C_o}}$$

C_o can be calculated as follows:

$$C_o = 1000\left(\frac{t_{pd}}{Z_o}\right) \text{ pF/length} \tag{4.12}$$

This loaded propagation delay value is one method that may be used to decide if a trace should be considered a transmission line ($2 \times t'_{pd} \times$ trace length $> t_r$ or t_f) where t_r is the rising edge of the signal and t_f is the falling edge.

C_d, the distributed capacitance per length of the trace depends on the capacitive load of the receiving devices, sockets, and vias. To mask transmission line effects, slower edge times are recommended. A heavily loaded trace slows the rise and fall times of the devices due to the increased RC time constant associated with the increased distributed capacitance and the filtering of high-frequency components out of the switching device. Heavily loaded traces seem advantageous until the loaded trace condition is examined. A high C_d raises the loaded propagation delay and lowers the loaded characteristic impedance. The higher loaded propagation delay value increases the likelihood that transmission line effects will not be masked during the rise and fall times, and a lower loaded characteristic impedance often exaggerates impedance mismatches between the driving device and the PCB trace. Thus, the apparent benefits of a heavily loaded trace are not realized unless the driving device is designed to drive large capacitive loads [3].

Loading also alters the characteristic impedance of the trace. As with the loaded propagation delay, a high ratio between the distributed capacitance and the intrinsic capacitance exaggerates the effects of loading on the characteristic impedance. Because $Z_o = \sqrt{L_o/C_o}$, the load adds capacitance. The loading factor $\sqrt{1 + C_d/C_o}$ divides in Z_0, and the characteristic impedance is lowered when the trace is loaded. Reflections on a loaded printed circuit trace, which cause ringing and stair-stepped switching delays, are more extreme when the loaded characteristic impedance differs substantially from the driving device's output impedance and the receiving device's input impedance [3].

If capacitive input loading is too high, compensating the clock is not practical. For this situation, use a series resistor to prevent reflection or ringing that may exist on the trace. Calculate the resistor value to be $Z_o - Z_d$, where Z_o is the clock line impedance and Z_d is the equivalent input impedance of all devices in the clock circuit trace.

The low impedance often encountered in the PCB sometimes prevents proper Z_0 impedance termination. If this condition exists, put as large as possible a series resistor in the trace (without corrupting signal integrity). Even a 10 Ω resistor is helpful; however, 33 Ω is commonly used.

It is important to note that the higher the switching speed (edge rate of the signal), the more important the series termination resistor from the clock driver must equal the value of Z_0. Typical values of the series termination resistor are 10 to 33 Ω

4.6 DECOUPLING

Clock circuit components must be RF decoupled with capacitors. This is due to the switching energy generated by the component injected into the power and ground planes. This energy will be transferred to other circuits or subsections as common-mode and/or differential-mode RF noise. The subject of decoupling is presented in Chapter 3 and is briefly discussed in this section. Bulk capacitors such as tantalum and high-frequency ceramic monolithic are both required, each for a different application. Furthermore, monolithic capacitors must have a self-resonant frequency higher than the clock harmonics requiring suppression. Typically, one selects a capacitor with a self-resonant frequency in the range of 10 to 30 MHz for circuits with edge rates of 2 ns or less. Many PCBs are self-resonant in the 200 to 400 MHz range. Proper selection of decoupling capacitors, along with the self-resonant frequency of the PCB (acting as one large capacitor) will provide enhanced EMI suppression. Tables 3.1 and 3.2, located in Chapter 3, are useful for axial or radial lead capacitors. Surface mount devices have a much higher self-resonant frequency by two orders of magnitude (or 100×) due to lower lead length inductance. Aluminum electrolytic capacitors are ineffective for high-frequency decoupling and are best suited for power supply subsystems or power line filtering.

Always provide adequate high-frequency RF decoupling capacitors in addition to bypass capacitors in all clock circuit areas. Calculate the capacitance required to suppress RF energy for all significant clock harmonics. Choose a capacitor with a self-resonant frequency higher than the clock harmonics requiring suppression, generally considered to be the fifth harmonic of the original clock frequency. Calculate capacitive reactance (self-resonant frequency in ohms) of decoupling capacitors per Eq. (4.13).

$$X_c = \frac{1}{2\pi fC} \tag{4.13}$$

where

X_c = capacitance reactance (ohms)

f = resonant frequency

C = capacitance value

4.7 TRACE LENGTHS

When placing PCB components during layout that use clocks or periodic signals, locate these devices so that clock traces are routed for a best straight-line path possible with minimal trace length and number of vias. Vias add inductance to the trace (approximately 1–3 nH each). Inductance in a trace may cause functional signal quality concerns and potential RF emissions. The faster the edge rate of the clock signal, the more this design rule approaches mandatory status. If a periodic signal or clock trace must traverse from one routing plane to another, this transition should occur at a component lead (pin escape) and not anywhere else, if possible, to reduce additional inductance presented to the trace from a minimum of two extra vias.

Table 4.3 illustrates different impedance values of a trace 3.2 mm (1/8") wide. These numbers are different from Table 4.1 due to the width (W) of the trace.

Any periodic signal or clock circuitry located within 2 inches (5 cm) of I/O components (or I/O connectors) should have edges (t_r) slower than 10 ns, since most I/O circuits (serial, parallel, audio, and the like) are generally slow compared to other functional sections. Traces located within 3 inches (7.6 cm) of the I/O section should have an edge rate between 5

Table 4.3 Impedance of a Trace 32 mm (1/8") Wide

Frequency (MHz)	*Impedance (Ω) at Various Lengths*		
	1"	*3"*	*10"*
1	0.13	0.38	1.25
10	1.25	3.75	12.5
100	12.5	37.5	125.0

and 10 ns. This general rule of locating clocks near I/O areas is not required when functional partitioning occurs (see Chapter 5). This is because functional partitioning of the power and ground planes in an I/O area will prevent RF currents that exist in other sections of the board from entering the I/O area. These RF currents can be coupled onto I/O cables and radiated to the external environment as common-mode or differential-mode noise. Keeping RF currents created from periodic or clock signals entering I/O circuitry is the ultimate design objective.

The old directive to "keep all clock lines short" is valid. The longer the trace, the greater the probability that RF currents will be produced and more spectral distribution of RF energy created. Clock traces must be terminated to reduce ringing and creation of avoidable RF currents. This is because unterminated lines generate signal reflections that can cause EMI to be generated. Detailed discussion is presented in the next section. Clock traces might also be degraded to the point of being nonfunctional due to ringing.

Ringing is caused by reflections and mismatches in the trace. These reflections can corrupt signal quality (and possibly functionality) of the circuit. They either add or subtract (phasing of the signal), depending on the net result of phasing. These signals (if added) will degrade to the point where they become an invalid clock. Use of transmission line theory allows signals to travel between source and load without creating reflections.

4.8 IMPEDANCE MATCHING—REFLECTIONS

As signal edges become faster, consideration must be given to the propagation and reflection delays of the routed trace. If the propagation time of the trace is longer than the edge time from the source to load, an electrically long trace will exist. This electrically long trace can also cause possible signal functionality problems (depending on the type and nature of the signal). This includes crosstalk, ringing, and reflections. EMI concerns are usually secondary to signal quality when referenced to electrically long lines. Although these long traces may exhibit oscillations, other suppression and containment measures implemented within the product may mask the EMI energy created. As a result, the device may cease to function properly if impedance mismatches exist in the system between source and load. Reflections are frequently a signal quality and EMC problem when the edge time of the signals constitutes a significant percentage of the propagation time between the device load intervals. Solutions to reflection problems often require extending the edge time or decreasing the distance between the load device intervals, or using point-to-point radials with correct termination.

Unterminated signal traces must be terminated in a matched load to avoid pulse ringing and match phase relationships. Figure 4.2 shows different types of signal reflections that may exist. Detailed analysis of transmission line characteristics and theory is available in the sources listed in the Bibliography and is beyond the scope of this design guide.

Examine Fig. 4.2 in detail. When a properly terminated transmission line exists, a smooth signal pulse is propagated down the trace from source to load. This is shown in Fig. 4.2*a*. No overshoot, undershoot, or reflections are observed. Theoretically, this is the best case signal possible. In reality, this 100 percent clean signal can exist only under ideal design conditions. Real components will always have some ringing generated by the output switching transistors, which are non-ideal devices in the first place.

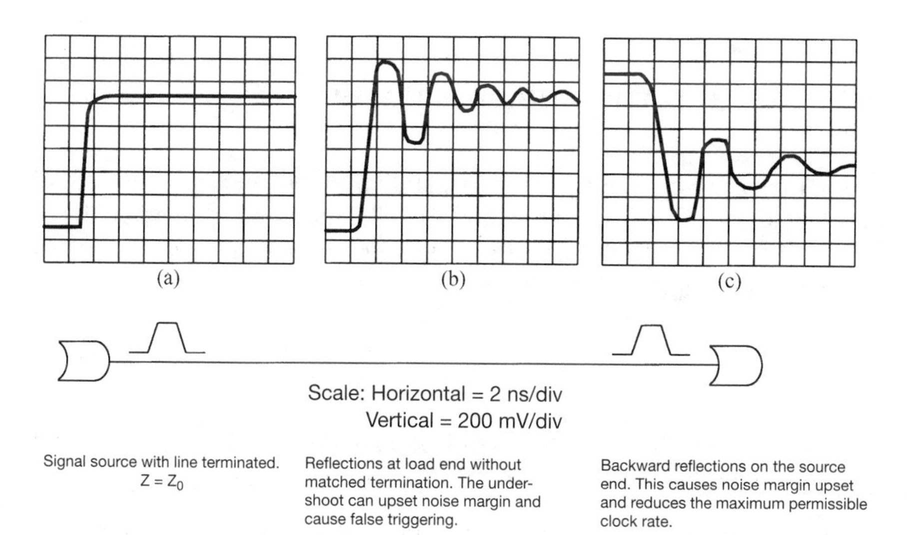

Fig. 4.2 Ringing on traces (Source: reprinted from Ref. 1 by permission)

In Fig. 4.2*b*, the input gate of a load with an electrically long signal trace is shown. Here it is observed that ringing (overshoot and undershoot) exist. Undershoot, if severe enough, can falsely trigger the load into believing that a logic "0" state is present. This false triggering then causes improper operation of the circuit. If the length of the trace is very long with respect to the propagation delay of the signal (source-to-load and load-to-source), reflections will be created that bounce back and forth between these two points. Depending on the frequency of the clock signal and edge rate of the source driver, phasing of the signal will be either additive or subtractive. If additive, signal functionality concerns exist, and the result can be as drastic as total system failure. If subtractive, the ringing will be damped to an acceptable level of operation. Correct termination can absorb this unwanted energy that causes the ringing.

Looking at Fig. 4.2*c*, observe what happens at the source of an unterminated line. Backward reflections can cause noise margin upset and corrupt the quality of the desired signal. These reflections are also caused by an electrically long signal trace as described above for Fig. 4.2*c*. When backward reflections exist, the edge rate desired for proper operation is reduced to a slower time period. This clock degradation may be sufficient enough to cause other functional sections of the PCB from clocking at the intended speed of operation; hence, performance is degraded to that of nonfunctionality.

What determines an electrically long trace? This can easily be calculated as described next.

4.9 CALCULATING TRACE LENGTHS

Assuming a typical velocity of propagation that is 60 percent of the speed of light, calculate the maximum permissible unterminated line length per Eq. (4.14). This equation is valid when the two-way propagation delay (source-load-source) equals the signal rise time.

$$L_{max} = \frac{t_r}{2t'_{pd}} \tag{4.14}$$

where t'_{pd} = the propagation delay, t_r in nanoseconds and L_{max} in centimeters. Figure 4.3 illustrates this equation for quick reference.

To simplify Eq. (4.14), and using Eqs. (4.4) and (4.8) (factoring in propagation delay and constant), Eqs. (4.15) and (4.16) are presented for

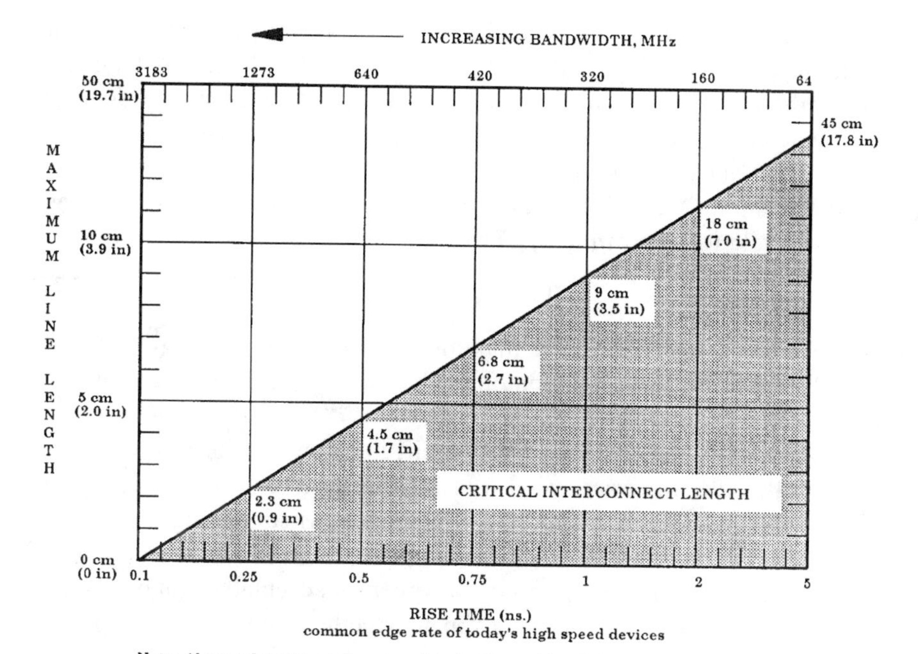

Fig. 4.3 Maximum unterminated line length vs. signal edge rate

determining the maximum electrical line length before termination is required.

$$L_{max} = 9 \times t_r \qquad \text{(for microstrip topology, in cm)} \qquad (4.15)$$

$$L_{max} = 7 \times t_r \qquad \text{(for stripline topology, in cm)} \qquad (4.16)$$

For example, if a signal edge is 2 ns, the maximum unterminated trace length when routed microstrip is:

$$L_{max} = 9 \times t_r = 18 \text{ cm } (7")$$

When this same clock trace is routed stripline, the maximum unterminated trace length of this 2 ns signal edge becomes:

$$L_{max} = 7 \times t_r = 14 \text{ cm } (5.5")$$

If a trace is longer than L_{max}, then termination (at the load) should be implemented as signal reflections (ringing) may now exist in this electrically long trace. Termination of traces is presented in Section 4.13. Ringing generated by an impedance mismatch in an electrically long trace may also make the circuit nonfunctional and could create RF currents. Even with good termination, RF currents will still exist on the trace. Use of a series resistor (source location) will minimize the RF current on the trace, prevent reflections (ringing), match trace impedance, minimize overshoot and undershoot, and reduce RF energy generated by slowing down edge rates of the clock signal.

The designer should calculate trace length to minimize reflections using the relationship of Eq. (4.17):

$$L_d < L_{max} \tag{4.17}$$

where L_{max} is the calculated maximum trace length [Eqs. (4.15) and (4.16)], and L_d is the actual length of the trace route as measured in the actual board layout.

If a clock trace must be electrically long and contain clock edges faster then 3 ns, route this trace using transmission line techniques. Doing so minimizes ringing and reflections in addition to maintaining a constant impedance on the trace. Ideally, trace impedance should be kept between ±10 percent. In some cases, ±20 to 30 percent is acceptable, but only after careful consideration has been given to whether impedance controlled traces, if any, will be affected by this broader tolerance level. The width of the trace, its height above an image plane, dielectric constant of the board material, plus other microstrip and stripline parameters determine the impedance of the trace [Eqs. (4.2) to (4.8)]. Maintain constant impedance control at all times, especially for periodic or clock signal traces.

Another mathematical method used to determine termination of a signal trace using characteristic impedance, propagation delay, and capacitive loading equations is now presented, first for microstrip and then for stripline [3].

4.9.1 Microstrip

A 5 ns edge rate device is provided on a 5 inch surface microstrip trace. There are six loads (logic devices) distributed throughout the route. Each device has an input capacitance of 6 pF each. Is termination required for this route?

Geometry

Trace width, W = 0.010 inch

Height above a plane, H = 0.012 inch

Trace thickness, T = 0.002 inch

Dielectric constant, $E_r = 4.7$

A. Calculate the characteristic impedance and propagation delay using Eqs. (4.2) and (4.4).

$$Z_o = \left(\frac{87}{E_r + 1.41}\right)\ln\left(\frac{5.98 \times H}{0.8 \times W + T}\right)$$

$$= \left(\frac{87}{\sqrt{4.7 + 1.41}}\right)\ln\left(\frac{5.98 \times 12}{0.8 \times 10 + 2}\right) = 69.4\ \Omega$$

$$t_{pd} = 1.017\sqrt{0.475E_r + 0.67} = 1.73\text{ ns/ft }(0.144\text{ ns/in})$$

B. Analyze capacitive loading.

Calculate C_d, distributed capacitance (total input capacitance divided by length)

$$C_d = 6 \times C_d/\text{trace length} = (6 \times 6\text{ pF})/5\text{ in} = 7.2\text{ pF/in}$$

Calculate intrinsic capacitance of the trace [Eq. (4.12)].

$$C_o = t_{pd}/Z_o = 1.73/69.4 = 2.08\text{ pF/in}$$

Calculate one-way propagation delay time from the source driver [Eq. (4.9)].

$$t_{pd}' = t_{pd}\sqrt{1 + \frac{C_d}{C_o}} = 0.144\sqrt{1 + \frac{7.2}{2.08}} = 0.30\text{ ns/in }(3.65\text{ ns/ft})$$

C. Perform a transmission line analysis.

Ringing and reflections are masked during edge transitions if

$$2 \times t_{pd}' \times \text{trace length} \le t_r \text{ or } t_f$$

For this situation,

$$2 \times t_{pd}' \times \text{trace length} = 2 \times 0.30\text{ ns/in} \times 5\text{ in} = 3.00\text{ ns}$$

Given that the edge rate of the component is $t_r = t_f = 5$ ns, and propagation delay ($3.00 \le 5$), termination is not required. Sometimes, the guideline of ($3 \times t_{pd} \times$ trace length) is used to provide a margin of safety. For this case, propagation delay would be 4.5 ns; hence, termination would still not be needed.

Assume now that the trace is routed stripline. Is termination required? From above:

$$t_{pd} = 1.017 \times \sqrt{E_r} = 2.20\text{ ns/ft} = 0.18\text{ ns/in}$$

$$C_o = t_{pd}/Z_o = 0.18/69.4 = 2.59\text{ pF/in }(35\text{ pF/ft})$$

$$C_d = \text{same as above } (7.2 \text{ pF/in})$$

$$t_{pd}' = t_{pd}\sqrt{1 + C_d/C_o} = 0.35 \text{ ns/in } (4.28 \text{ ns/ft})$$

$$2 \times t_{pd}' \times \text{trace length} = 2 \times 0.35 \text{ ns/in} \times 5 \text{ in} = 3.50 \text{ ns}$$

Again, this trace would not require termination since 3.50 ns ≤ 5 ns. Propagation delay for stripline is 0.50 ns longer. This is due to the fact that t_{pd} (unloaded) is substantially greater than microstrip. This factor helps prevent the transmission line effects from being masked during edge rate changes.

4.9.2 Loaded Stripline

A 2 ns edge rate device on a 10 inch stripline trace is used. There are five logic devices distributed throughout the route. Each device has an input capacitance of 12 pF. Is termination required for this route?

Geometry

Trace width, W = 0.006 inch
Distance from a plane, H = 0.020 inch
Trace thickness, T = 0.0014 inch
Dielectric constant, $E_r = 4.6$

A. Calculate characteristic impedance and propagation delay [use Eqs. (4.6) and (4.8)].

$$Z_o = \left(\frac{60}{\sqrt{E_r}}\right)\ln\left[\frac{4 \times H}{0.67\pi W \times \left(0.8 + \frac{T}{W}\right)}\right]$$

$$= \left(\frac{60}{\sqrt{4.6}}\right)\ln\left[\frac{4 \times 20}{0.67\pi 6 \times \left(0.8 + \frac{1.4}{6}\right)}\right] = 50.7\ \Omega$$

$$t_{pd} = 1.017\sqrt{E_r} = 2.18 \text{ ns/ft } (0.182 \text{ ns/in})$$

B. Analyze capacitive loading.
Calculate C_d, distributed capacitance (total input capacitance divided by length).

$$C_d = 6 \times C_d/\text{trace length} = (6 \times 12 \text{ pF}) / 10 \text{ in} = 7.2 \text{ pF/in}$$

Calculate the intrinsic capacitance of the trace.

$$C_o = t_{pd} / Z_o = 0.182 / 50.7 = 3.58 \text{ pF/in } (43.0 \text{ pF/ft})$$

Calculate the one-way propagation delay time from the source driver.

$$t_{pd}' = t_{pd}\sqrt{1 + 7.2/3.58} = 0.32 \text{ ns/in } (3.79 \text{ ns/ft})$$

C. Perform transmission line analysis.
The important item of interest is $2 \times t_{pd}' \times$ trace length $\leq t_r$ or t_f.

$$2 \times t_{pd}' \times \text{trace length} = 2 \times 0.32 \text{ ns/in} \times 10 \text{ in} = 6.32 \text{ ns}$$

Since the edge rate of the component $t_r = t_f = 2$ ns, and propagation delay is ($6.32 \geq 2$), termination is required to absorb transmission line effects.

Now assume the trace is routed microstrip. Is termination required? From above,

$$t_{pd} = 1.017 \times \sqrt{0.475E_r + 0.67} = 0.14 \text{ ns/in } (1.72 \text{ ns/ft})$$

$$C_o = t_{pd}/Z_o = 0.14/50.7 = 2.76 \text{ pF/in } (33 \text{ pF/ft})$$

$$C_d = \text{same as above } (7.2 \text{ pF/in})$$

$$t_{pd}' = t_{pd} \sqrt{1 + C_d/C_o} = 0.26 \text{ ns/in } (3.19 \text{ ns/ft})$$

and

$$2 \times t_{pd}' \times \text{trace length} = 2 \times 0.26 \text{ ns/ft} \times 10 \text{ in} = 5.20 \text{ ns}$$

Again, this trace would require termination since 5.20 ns $\geq$ 2 ns.

4.10 ROUTING LAYERS

Designers frequently overlook the question of which layers or planes to use for routing clocks and periodic signals. These *must be routed on either one plane only, or on an adjacent plane separated by a single image plane.* This is shown in Fig. 4.4. Three issues must be remembered when select-

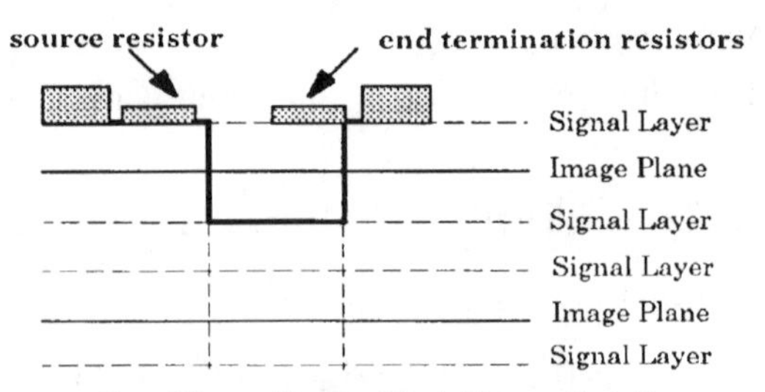

Best Way to Route Clock Traces Stripline
(total of 2 vias)
The perpendicular traces complement ach other

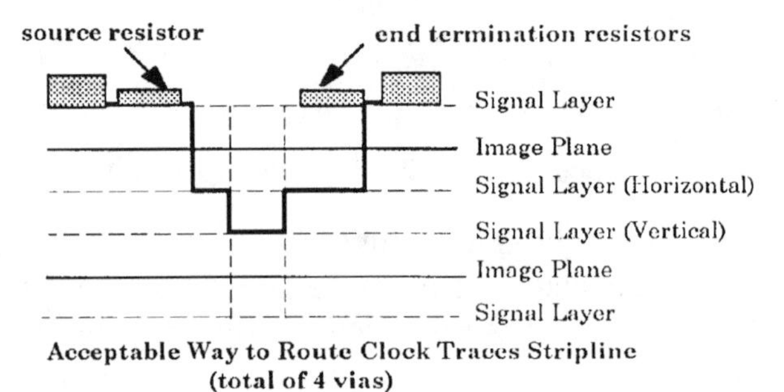

Acceptable Way to Route Clock Traces Stripline
(total of 4 vias)

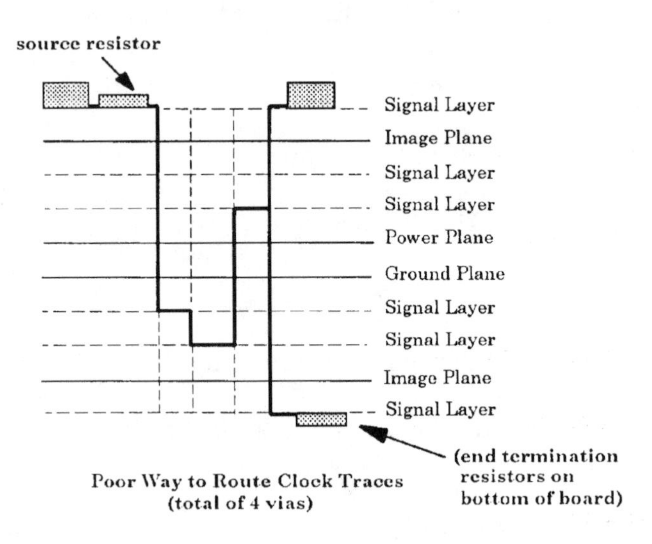

Poor Way to Route Clock Traces
(total of 4 vias)

Vias are represented by vertical dashed lines.

NOTE: **This method maximizes use of vias and layer jumping, both of which must be minimized.**

Fig. 4.4 Routing layers for clock signals

ing routing layers: (1) which layers to use for trace routing, (2) jumping between designated layers, and (3) constant trace impedance. These are discussed below.

4.10.1 Routing layers

Figure 4.4 is a representative sample of how to route clock traces.

1. The designer needs to use a solid (image) plane adjacent to the clock trace. Minimize trace length while maintaining controlled impedance. If a series termination resistor is used, connect the resistor to the pin of the component *without use of a via between the resistor and component.* Place the resistor on the top side of the board next to the output pin of the device. After the resistor, place a via to the internal stripline layers. The use of ground planes is preferred over use of voltage planes because ground planes provide for greater flux cancellation of RF currents.
2. Do not route clock traces on the bottom layer of a multilayer board (i.e., beneath the power and ground planes) if a six-layer or greater stackup is used.

   The bottom half of a board (below the center power/ground planes) is generally reserved for large signal buses and I/O circuitry. Functional signal quality of these traces could be corrupted by placing high-threat signals in this quiet area. When routing traces on lower levels, there exists a change in the distributed capacitance of the trace (as the trace relates to a plane), thus affecting performance and possible signal degradation.
3. If we maintain constant trace impedance and minimize or eliminate use of vias, the trace will not radiate any more than a coax. When we reference the electric field, **E**, to an image plane, magnetic flux is cancelled.

Three phenomena by which planes (and, hence PCBs) create EMI are enumerated below. Proper understanding of these concepts will allow the designer to enhance suppression on *any* PCB.

1. discontinuities in the image plane due to use of vias and jumping clock traces between layers
2. peak surge currents created in the image plane due to components switching signal pins simultaneously (refer to Chapter 2)
3. flux loss into the annular keep-out region of the via if 3-W routing

is not provided for the trace (Distance separation of a trace from a via must also conform to 3-W spacing. The 3-W rule is discussed in section 4.16.)

The advantage and disadvantage of routing clock traces by microstrip and stripline techniques is shown in Figs. 4.5 and Fig. 4.6 and described below.

- *Outer layer, microstrip.* This routing allows for the fastest transition of signal edges as a result of less distributed capacitance between the trace and its closest image plane in addition to the propagation delay of the dielectric material. Capacitance rounds off or slows down clock edges. The negative aspect of routing traces on the outer layer (microstrip) is that any RF energy generated on the trace can radiate, thus causing potential noncompliance with EMI requirements. This is due to lack of flux cancellation of RF currents from the trace to the first plane. This lack of cancellation is caused by return flux eddys around vias and potential cross coupling to and from other signals on adjacent traces.

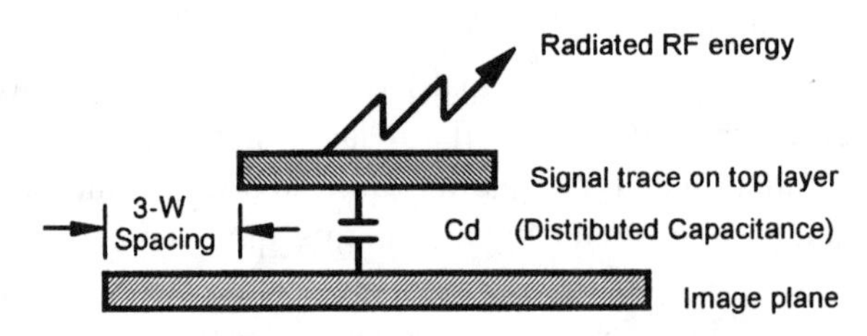

Fig. 4.5 Routing clock traces microstrip (layer 1)

- *Inner layers, stripline.* This routing topology is optimal for suppression of common-mode RF currents, which is the result of the effect of the adjacent image planes. Routing clocks stripline could also cause increased signal propagation time due to additional distributed capacitance present in the planes. Although this distributed capacitance is minimal, clocks with edges in the nanosecond and sub-nanosecond range may be skewed by this capacitive effect. An advantage of routing stripline is that most RF energy generated by traces is captured by the two image planes, thus keeping this RF energy internal to the PCB instead of radiating to the external environment or coupling to other circuits.

  Note, however, that component (circuit device) radiation is not eliminated. When using stripline, components are still outside the envelope of protection provided by both planes. In most applications, component occupancy areas dominate in providing a source of RF energy. This is certainly true of through-hole devices. This component-specific radiation can be significant.

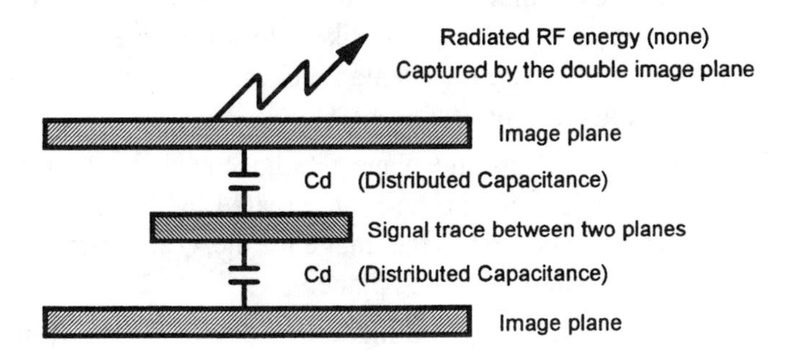

Fig. 4.6 Routing clock traces stripline

4.10.2 Layer jumping—use of vias

When routing clock or high-threat signals stripline, it is common practice to via the trace to a routing plane (e.g., x-axis) and then via this same trace to another plane (e.g., y-axis) from source to load. This is shown in the poor routing method of Fig. 4.4. It is generally assumed that if each and every trace is routed adjacent to an image plane, there will exist tight RF coupling of common-mode RF currents along the entire route. In reality, this assumption is partially incorrect.

Refer to the section on image planes and Fig. 2.22 (Chapter 2). Notice that, as the signal trace jumps from one layer to another, the RF return current should follow the trace route. When using image planes, RF currents will return in the path of least impedance. In a stripline configuration, the return current is shared between two image planes adjacent to the horizontal and vertical routing layers.

When a jump is made from a horizontal to a vertical layer, the RF return current *cannot* make this jump. This is because a discontinuity is placed in the trace route by the via. The return current must now find an alternative low-inductance (impedance) path to complete its route. This alternative path may not exist in a position that is immediately adjacent to the location of the via used for the jump. As a result, RF currents on the signal trace can couple to other circuits and pose problems as both crosstalk and EMI. Use of vias in a route will always create a concern in any high-speed, high-technology product.

To minimize creation of EMI and crosstalk due to layer jumping, the following design techniques have been found effective:*

- Route all clock and high-threat signal traces on only one routing layer. This means that both x- and y-axis routes are in the same plane. (Note: This technique is likely to be rejected by the CAD layout person as being unacceptable because it makes autorouting of the board nearly impossible.)
- Verify that a solid ground plane is adjacent to the routing layer, with no discontinuities in the route created by use of vias, where the ground plane is used as the image for the x- and y-axis immediately above or below the same plane.
- If a via must be used for routing a sensitive (high-threat or clock) signal trace between the horizontal and vertical routing layer, the designer should incorporate adjacent ground vias at "each and every" via location where the axis jumps are executed.

A ground via is a via that is placed directly adjacent to each signal route via. Ground vias are used when there is more than one ground plane in the PCB. This via is connected to the ground planes in the board that serve as the image planes for the signal jump currents. What this via does is essentially tie the ground planes together adjacent to this signal trace via location. When using two ground vias per signal trace via, a continuous image plane for the trace to couple its RF return current throughout its entire route (a ground plane built into the via) exists. This ground via will maintain a constant image plane located adjacent to a signal route, in the vertical axis.

4.11 GUARD/SHUNT TRACES

Guard traces may be used to surround clocks, periodic signals, differential pairs, or system critical (high-threat) traces from source to destination. Shunt traces are traces located *immediately* above a high-threat signal trace that traverses the PCB along its entire route. Both guard and shunt traces have unique applications, implementations, and drawbacks. Depending on functional requirements, one or both techniques may be used. It is up to the design engineer to select which or both techniques are required for EMI suppression. Guard traces are particularly useful to contain signal loop areas in two-layer boards. In microstripline or stripline

* Use of ground vias was first identified and presented to industry by W. Michael King.

boards, following the 3-W rule usually will provide an adequate flux boundary, circumventing the usefulness of guard traces.

A PCB with a solid ground plane can produce common-mode RF currents all by itself. This is because RF current encounters a finite inductance (impedance) in the ground plane material (usually copper). This inductance creates a voltage gradient, commonly called *ground-noise voltage.* This voltage (also termed *ground shift*) and an equivalent shift in the power planes, is responsible for a very significant amount of common-mode EMI in circuit boards. This voltage gradient causes a small portion of the signal trace current to flow through the distributed stray capacitance of the ground plane. This is illustrated in Fig. 4.7, where the following abbreviations are used:

L_s = partial self-inductance of the signal trace

M_{sg} = partial mutual inductance between signal trace and ground plane

L_g = partial self-inductance of the ground plane

M_{gs} = partial mutual inductance between ground plane and signal trace

C_{stray} = distributed stray capacitance of the ground plane

V_{gnd} = ground plane noise voltage

To calculate ground-noise voltage, V_{gnd}, we can use Eq. (4.18).

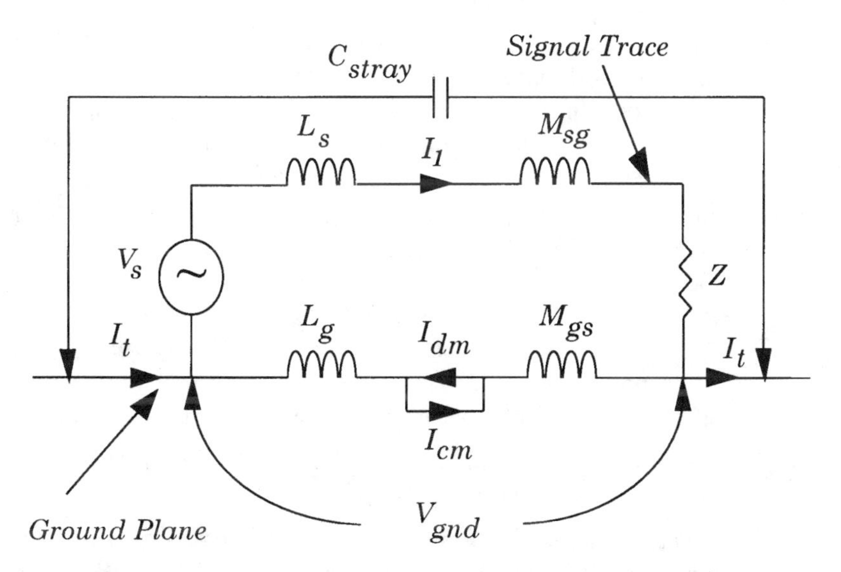

Fig. 4.7 Schematic representation of a ground plane (Source: reprinted from Ref. 4 by permission.)

$$V_{gnd} = L_g \frac{dI_{dm}}{dt} - M_{gs} \frac{dI_l}{dt} \quad (4.18)$$

To reduce the total ground-noise voltage, increase the mutual inductance between the trace and its nearest image plane. Doing so provides an additional path for signal return currents to image back to their sources.

Generally, common-mode currents are typically several orders of magnitude lower than differential-mode currents. It should be remembered that the common-mode current is a by-product of the differential-mode switching that does not become cancelled. However, common-mode currents (I_l and I_{cm}) produce higher emissions than those created by differential-mode (I_l and I_{dm}) currents. This is because common-mode RF current fields are additive, whereas differential-mode current fields tend to cancel. This is best illustrated by Fig. 2.19, in Chapter 2.

To reduce I_t currents, the ground-noise voltage must be reduced. This is best accomplished by reducing the distance spacing between the signal trace and ground plane. In most cases, this is not fully possible, because the spacing between a signal plane and image plane must be at a specific distance to maintain a constant impedance of the board, as demonstrated by Eqs. (4.2) through (4.8). Hence, there are prudent limits to distance separation between the two planes. The ground noise voltage must still be reduced. This ground-noise voltage can be reduced by providing additional paths for RF currents to flow [4]. These additional paths are provided by guard and shunt traces.

Guard and shunt traces are used to provide an additional return path for common-mode currents. Guard traces will only provide an additional return path if placed adjacent to the high-threat and high-impedance trace with respect to the image plane. This distance from signal trace to guard traces must be as close as can be manufactured. Consider guard and shunt trace techniques as appropriate for providing this alternative return path in several application-dependent ways. The advantages of guard and shunt traces include the following:

1. *To enforce the 3-W rule (discussed later in this chapter).* This assists in prevention of crosstalk that might occur between high threat signal traces and other nearby components or other traces.
2. *To prevent common-mode RF coupling from high-threat signal traces (small line width) to other circuit traces.*
3. *To provide an additional low-impedance "alternative return path" and minimize RF common-mode currents that exist in the image plane.* This is observed more with shunt than with guard traces.

4. *To create an impedance controlled, coaxial based transmission line effect.* This is shown in Fig. 4.8, where both guard and shunt traces are used.

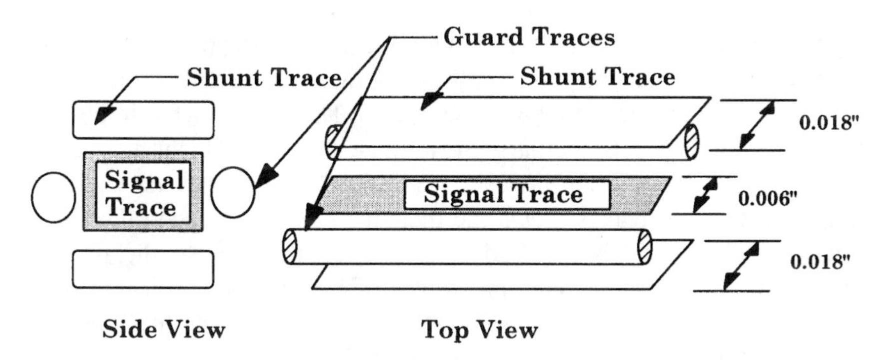

Fig. 4.8 Shunt traces

Guard traces perform best on one- and two-layer boards (i.e., those without a power or ground plane). RF coupling occurs between a signal trace and a solid plane through both capacitive and inductive means (including mutual inductance) to remove common-mode currents and contain signal-to-return loop areas. Image planes provide tight RF coupling for these return currents. Without a solid plane on a one- or two-layer board, a reference trace (return path for RF currents) must be provided in some manner.

Shunt traces perform best in multilayer boards (six or more layers). Shunt traces sandwich a high-threat signal trace between two separate ground reference sources (image plane and shunt trace). The advantage of using a shunt trace results from the skin effect of the currents flowing on the copper trace. There is no significant current flow inside the center of a trace. A detailed discussion on this subject is found in Chapter 2. Placing a shunt trace above and/or below the signal trace provides additional coupling to the trace as observed in Fig. 4.8.

Note that the guard traces in Fig. 4.8 are shown as circular traces. This is to distinguish them from shunt traces. In reality, these traces are flat, like the signal and shunt traces.

In Fig. 4.8, visualize the assembly as if looking down a tube—a signal trace surrounded on top and bottom by a shunt trace (or plane) at ground potential. When combined with guard bands, there exists a configuration that resembles a coaxial transmission line. A coaxial transmission line maintains a constant impedance throughout its route. In addition, signal quality is enhanced while preventing common-mode currents created within the trace from causing EMI problems with adjacent circuits.

If crosstalk is a concern and emissions are not, a guard trace finds excellent use. This guard trace enforces the 3-W rule for placement of periodic or clock traces in relation to adjacent signal traces and low-level analog devices. Thus, RF radiated coupling is contained.

When a shunt trace is provided in a multilayer board, place it immediately vertically adjacent to the high-threat signal trace and connect *both* ends of the shunt trace to the ground planes. The shunt traces should not have voids in them—particularly those caused by vias. This is applicable "only" when two stripline layers are adjacent. Ensure that the width of the shunt trace is at least *twice* the distance spacing between the shunt trace and the signal trace. Additional via connections to the ground planes removes possible standing waves of RF currents projected onto the shunt trace [4]. Implementation of this scheme is rarely feasible, and impossible in many cases, so it is presented here for discussion only.

When shunt traces are grounded, an interesting phenomenon is observed. An LC resonance occurs (L from the trace and C from the distributed capacitance between trace and plane). Depending on the distance spacing between ground connections, a sharp resonant impedance may occur. If any harmonic of the clock signal lands at this resonant frequency, suppression of RF currents is made more difficult, and the board will radiate like a transmitter. Should this occur, install additional ground connections between the ground planes and shunt trace. Alter the distance spacing to shift this resonant frequency away from the clock harmonics.

If guard traces are used, minimize the spacing between the guard and signal trace to the smallest manufacturable spacing. This spacing must be maintained throughout the length of the route. The capacitive contribution of this spacing is minimal; however, suppression of harmonics could be significant. This is because the aspect ratio of the signal trace relative to the height above the ground plane is much less than the aspect ratio for coupling RF currents from the high-threat trace to a guard trace.

Ground the guard trace at both source and destination as observed in Fig. 4.9. Make this ground connection to the source driving the trace as close as possible as well as next to the destination point. If the routing lengths of both the signal and guard trace are significant, ***multiple connections to any ground planes by vias, along the edges of the guard trace, are also required.***

Do not make these multiple ground connections from the guard trace to the ground planes symmetrical or at the same distance spacing. Doing so is an invitation to trouble, since the length of the guard trace provides an alternative low-impedance path for common-mode RF currents to return

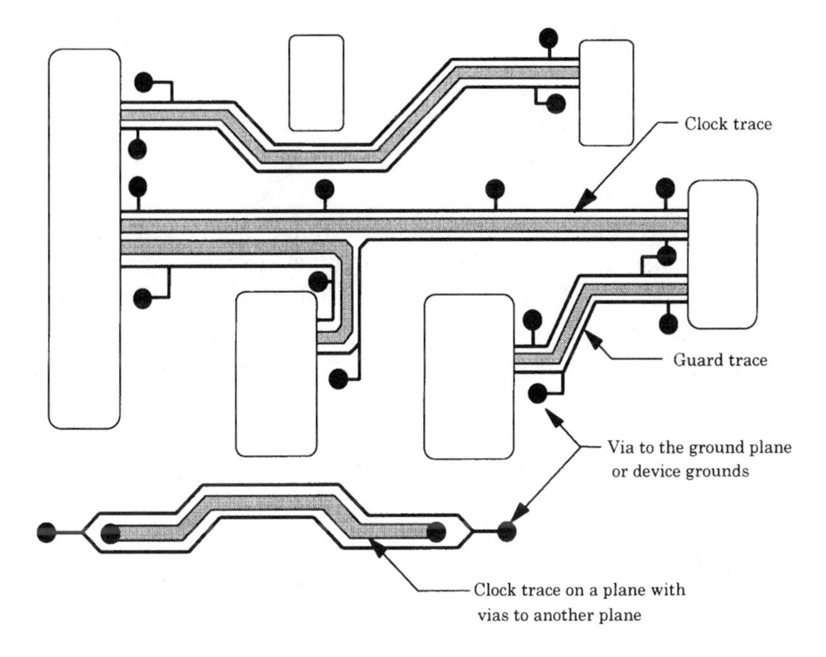

Fig. 4.9 Guard traces

to their source when referenced to an image plane. The LdI/dt currents, if at a multiple wavelength of the clock harmonics, could cause a resonance to occur (depending on the frequency and edge rate) that would intermodulate the trace and cause functional problems or the occurrence of RF emissions. This statement is true only for long traces (i.e., electrically long). In most cases, traces less than $\lambda/20$ of the fundamental wavelength are generally not affected by common-mode RF currents between trace and ground planes.

When a guard trace is forced away from a signal trace due to vias or through-hole component leads in the routing path, return this trace back to normal as soon as the detour is cleared. Never place anything between a signal trace and its guard trace.

When two or more periodic signal or clock traces are routed side by side, they may "share" a common guard trace between them for only a short distance, as shown in Fig. 4.9. Try to prevent routing two traces within the same guard trace if possible. Exceptions do exist, such as differential or paired signals.

It must be restated here that "guard" traces are primarily effective on one- or two-layer circuit boards. On microstripline or stripline boards, the flux boundary provided by the 3-W rule will accomplish much of the effect provided by guard traces.

4.12 CROSSTALK

Crosstalk can exist between traces on a PCB. This undesirable feature is associated not only with clocks and periodic signals, but also with data, address, control, and I/O traces.

Crosstalk is generally considered to be a functionality concern (signal quality) that causes a disturbance between traces. In reality, crosstalk is a major contributor in the creation of EMI. High-speed traces, analog circuits, and other high-threat signals may be corrupted by crosstalk induced from other sources. Conversely, high-speed traces may couple into lower-speed traces or sensitive circuits, causing both EMI emission and functionality problems. These EMI critical circuits may, however, also unintentionally couple their RF energy to I/O circuits. This coupling may result in radiated or conducted EMI that exits the enclosure or causes functionality problems between circuits and subsystems. An illustration of crosstalk and its associated circuit can be observed in Fig. 4.10.

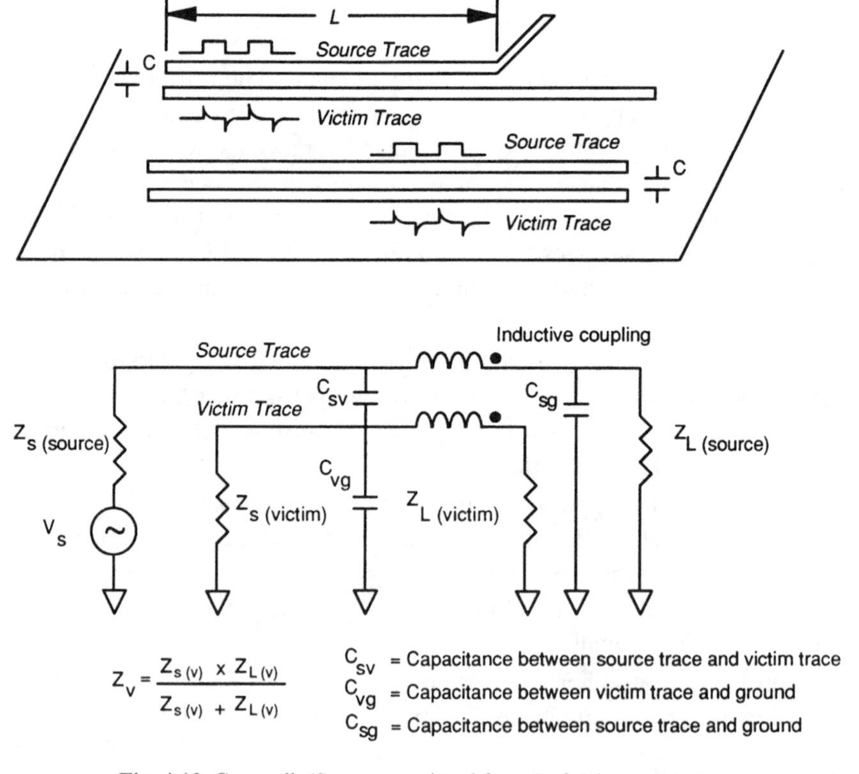

Fig. 4.10 Crosstalk (Source: reprinted from Ref. 1 by permission)

Crosstalk is measured in units of dBx. This is because the reference level is not an absolute power level. The reference is 90 dB loss from the interfering circuit to the victim circuit. As a result, this unit measures how much crosstalk coupling loss is above 90 dB. The relationship describing this is shown in Eq. (4.19).

$$\text{dBx} = 90 - (\text{crosstalk coupling loss in dB}) \tag{4.19}$$

For example, circuit A couples with circuit B. Circuit A is at a 58 dB lower power level. Calculate the crosstalk from circuit A to B as 32 dBx.

Crosstalk is also expressed by Eq. (4.20) when applied to a source and victim circuit.

$$X_{\text{talk (dBx)}} = 20\log\frac{V_{\text{victim}}}{V_{\text{source}}} \tag{4.20}$$

Crosstalk occurs by mutual capacitance and inductance between parallel traces. One trace (source) induces a certain percentage of its RF voltage into the other trace (victim). As the traces approach each other, a larger level of crosstalk noise is generated. Crosstalk is also a function of increasing frequency (or faster edge rate of signals) and higher victim impedance.

Use of the 3-W rule will assist in complying with PCB design criteria without having to implement guard traces. If guard traces cannot be used on single- or double-sided PCBs, the 3-W rule becomes mandatory when using shunt traces. To minimize coupling between traces and signals, the 3-W rule states that *"the distance separation between traces must be 3 times the width of the trace as measured from centerline to centerline of the two adjacent traces."* The 3-W rule is discussed in detail later in this chapter.

Do not restrict use of the 3-W rule to periodic signals or clock traces. Differential pair traces (balanced, ECL, etc.) are prime candidates for 3-W. Power plane noise can couple into paired traces, causing data corruption through crosstalk. In I/O sections where parallel traces are routed with absence of copper on all planes, an alternative technique is to route 3-W. Absence of copper is discussed in detail in Chapter 5.

Another technique to prevent or minimize crosstalk between parallel traces is to improve trace routing by separating traces with a minimum parallel length. This technique is preferred for long clock signals and high-speed parallel bus structures. Separate the two parallel traces at 2 mils per inch of trace length throughout the entire route.

To determine the distance spacing for microstrip traces for eliminating crosstalk, use Table 4.4. When using Table 4.4, notice that the dimension specified for $R_{V(total)} \approx 100\ \Omega$ and 1 cm length. For different R_V values, apply the correction $20 \log [(R_V \times length_{cm})/100]$. Clamp at maximum -4 dB for no ground plane and -10 dB for W/h = 1. For stripline, add 4 dB to the value. For other values of $R_{V(total)}$, use the equation provided in Table 4.4 [1].

4.13 TRACE TERMINATION

Trace termination plays an important role in reduction of RF energy. In addition to ensuring optimal signal functionality and quality, it enhances suppression. To prevent Z_0 corruption and provide higher-quality signal transfer between circuits, termination is required.

Engineers and designers will sometimes daisy-chain periodic signal and clock traces for easy routing. Unless the distance is small between loads (with respect to propagation length of the signal rise time), reflections may occur from daisy-chained traces. Daisy-chaining impacts signal quality and EMI spectral energy distribution, sometimes to the point of nonfunctionality or noncompliance. Therefore, radial connections for fast edge signals and clocks are preferred over daisy-chains. Each component should have its respective trace terminated in its characteristic impedance as shown in Fig. 4.11. Parallel termination at the end of the trace route is feasible when the drivers can tolerate the total current sink of the terminated loads. An excellent discussion on traces using transmission line theory can be found in Ref. 2, Chapters 3 and 7, and Ref. 3.

To prevent undesired effects of unterminated traces due to unmatched loads, termination may be required. There are five common termination methods available. These methods are dependent on the complexities of layout geometry, component count, power consumption, and other items detailed below. When a driving device is overloaded, termination can degrade the quality of the signal if incorrectly specified or implemented.

The five most commonly used termination methods are listed below and shown in Fig. 4.12 [2]. A summary of termination methods is detailed in Table 4.5.

1. series termination resistor
2. parallel termination resistor
3. Thevenin network
4. RC network
5. Diode network

Table 4.4 Capacitive Crosstalk Coupling Distance Spacing (for a PCB trace with $R_{v(total)} \approx 100\ \Omega$ and 1 cm length)

(Source: reprinted from ref. 1 by permission)

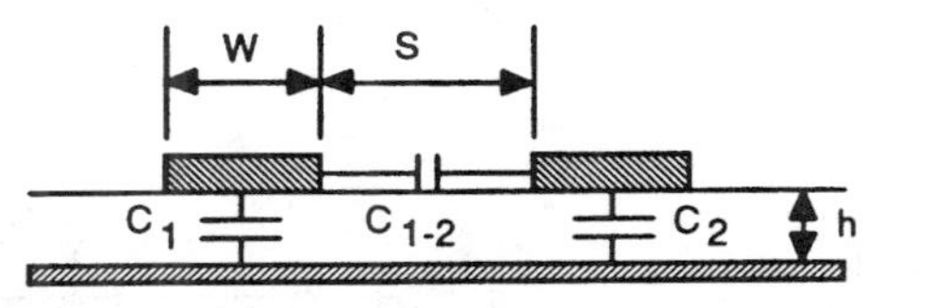

S/W (C_{1-2}, pF/cm)	W/h = 3 ($C1, C_2 \approx 1.2$ pF/cm) $Z_o = 50\ \Omega$				W/h = 1 ($C1, C_2 \approx 0.5$ pF/cm) $Z_o = 90\ \Omega$				W/h = 0.3 ($C1, C_2 \approx 0.1$ pF/cm) $Z_o = 120\ \Omega$			
	10 (0.003)	3 (0.02)	1 (0.06)	0.3 (0.17)	10 (0.003)	3 (0.02)	1 (0.06)	0.3 (0.17)	10 (0.003)	3 (0.02)	1 (0.06)	0.3 (0.17)
F= 1 kHz	-174	-158	-148	-140	-158	-148	-142	-136	-146	-134	-130	-122
3 kHz	-164	-148	-138	-130	-148	-138	-132	-126	-136	-124	-120	-112
10 kHz	-154	-138	-128	-120	-138	-128	-122	-116	-126	-114	-110	-102
30 kHz	-144	-128	-118	-110	-128	-118	-112	-106	-116	-104	-100	-92
100 kHz	-134	-118	-108	-100	-118	-108	-102	-96	-106	-94	-90	-82
300 kHz	-124	-108	-98	-90	-108	-98	-92	-86	-96	-84	-80	-72
1 MHz	-114	-98	-88	-80	-98	-88	-82	-76	-86	-74	-70	-62
3 MHz	-104	-88	-78	-70	-88	-78	-72	-66	-76	-64	-60	-52
10 MHz	-94	-78	-68	-60	-78	-68	-62	-56	-66	-54	-50	-42
30 MHz	-84	-68	-58	-50	-68	-58	-52	-46	-56	-44	-40	-32
100 MHz	-74	-58	-48	-40	-58	-48	-42	-36	-47	-34	-30	-22
300 MHz	-64	-48	-38	-30	-48	-38	-32	-26	-36	-24	-20	-12
1 GHz	-56	-40	-30	-22	-38	-30	-22	-18	-28	-18	-14	-8
3 GHz	-52	-36	-26	-20	-32	-24	-18	-14	-24	-14	-10	-4
10 GHz	-52	-36	-26	-20	-30	-22	-16	-10	-24	-14	-10	-4

$$\text{Crosstalk} = 20 \log \frac{R_{victim} C_1 - C_2 W}{\sqrt{[R_v W (C_{2-1})]^2 + 1}}$$

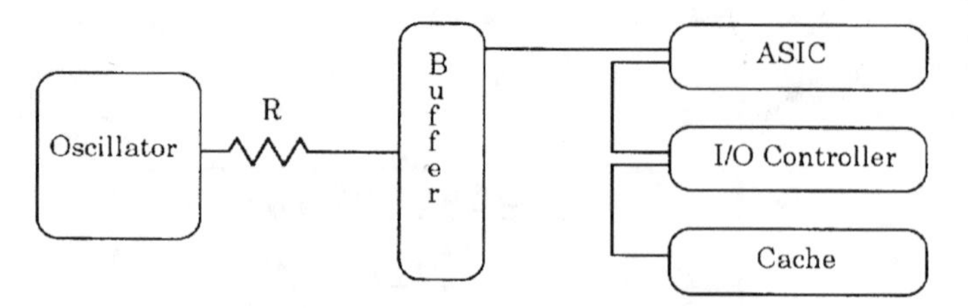

Poor Trace Routing for Clock Signals
(Note daisychaining of clock signal)

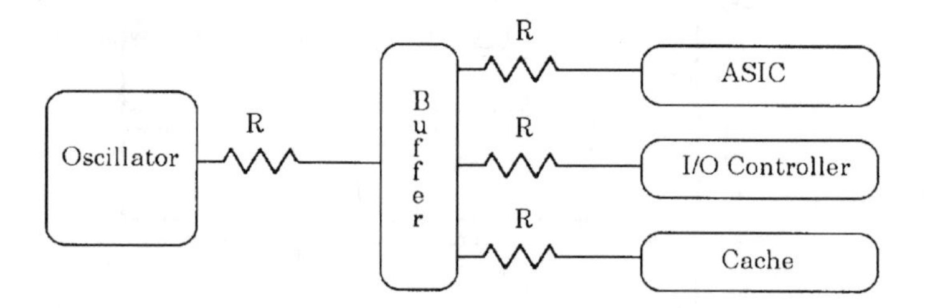

Optimal Trace Routing for Clock Signals
with Electrically Short Traces

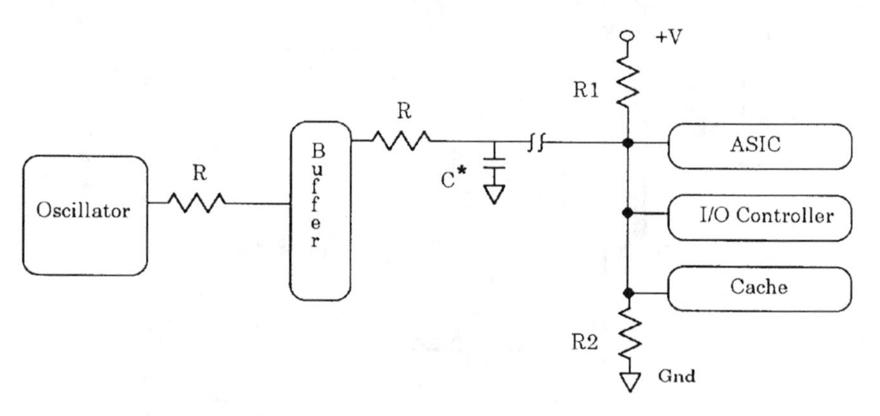

Optimal Trace Routing for Clock Signals
with Electrically Long Traces with respect to clock period

*** Capacitor optional - usually not needed with series termination**

R1 required if voltage drop in the trace is below acceptable levels of high transition state due to IR drop in the trace and number of load circuits

Fig. 4.11 Termination of clock traces

A combination of methods 1 and 2, or 1 and 3, is sometimes used with "electrically long" traces* where the series resistor is 15–75 Ω and end termination calculated for optimal performance.

Series termination resistor

Series termination is used when a lumped load is at the end of the trace. Use the resistor when the driving device's output impedance, Z_d, is less than Z_o, the loaded characteristic impedance of the trace, or when the fanout is low. Locate this resistor directly at the output of the driver, without use of a via between the component and resistor.

When the series resistor and Z_o are equal, the voltage waveform is divided evenly, with half of the voltage transmitted to the receiver. If the receiver has a very high input impedance, the full waveform will be observed immediately, while the source will receive the waveform at $2 \times t_{pd}$. Since devices have different input and output impedances that are not intuitively known, use of a series resistor may not be optimal.

Parallel termination resistor

For parallel termination, a single resistor is used. This resistor must have a Z_o value equal to the trace impedance which, in turn, should approximate the source impedance. The other end of the resistor is tied to a reference source, generally ground. The main disadvantage is that parallel termination consumes dc power, since its value is generally in the range of 50 to 150 Ω.

Table 4.5 Termination Types and Their Properties

Type	*Added Parts*	*Delay Added*	*Power Required*	*Parts Values*	*Comments*
Series	1	Yes	Low	$R_S = Z_o - R_d$	Good dc noise margin
Parallel	1	Small	High	$R = Z_o$	Power consumption is a problem.
Thevenin	2	Small	High	$R = 2 \times Z_o$	High power for CMOS
RC	2	Small	Medium	$R = Z_o$, $C = 300$ pF	Check bandwidth and added capacitance
Diode	2	Small	Low	—	Limits undershoot; some ringing at diodes

Source: reprinted from Ref. 3 by permission.

* An "electrically long" trace is one whose length in propagation time, source to termination, plus reflection time is equal to or longer than the rise time of the source.

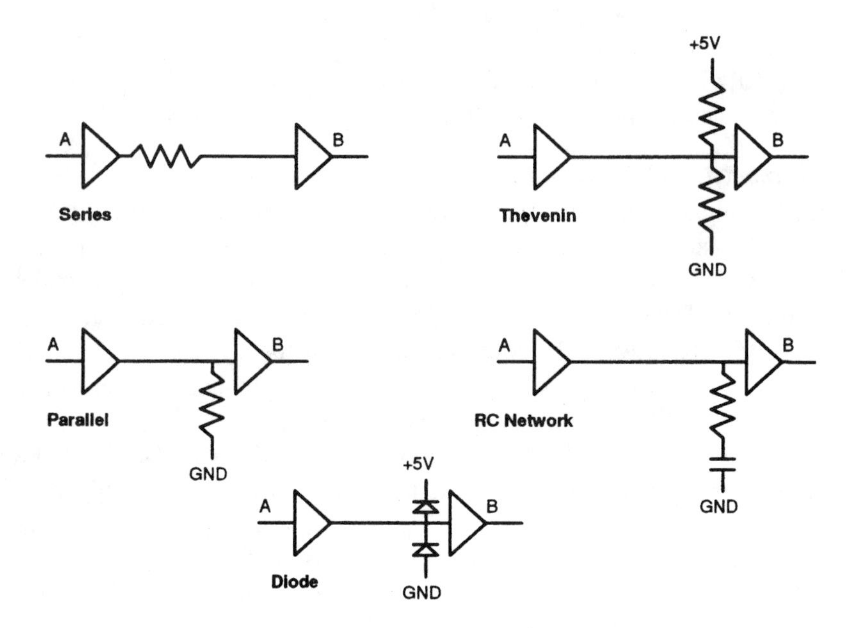

Fig. 4.12 Termination types (Source: reprinted from Ref. 3 by permission)

Thevenin network

This termination method connects one resistor to power and the other resistor to ground. It allows for proper transition points between logic high and low states. Thevenin termination is optimal for TTL logic. When used with CMOS components, keep in mind the switching voltage level related to the input voltage of the device. A poor choice of resistor values may cause the threshold levels to be altered, causing unreliable operation.

RC network

This termination method works well in both TTL and CMOS systems. The resistor matches the impedance of the trace. The capacitor holds the dc voltage level of the component. As a result, ac current flows to ground during the switching state. Although a minor delay is presented to the signal trace, less power dissipation exists than with regular parallel termination. The resistor must equal the Z_0 of the trace, while the capacitor is generally very small (20–600 pF). The RC time constant must be greater than twice the loaded propagation delay. RC termination finds excellent use in buses containing similar layouts.

Diode network

This termination method is commonly used on differential or paired networks. Diodes are used primarily to limit overshoot on traces while providing low power dissipation. The major disadvantage of diodes lies in their frequency response to high-speed signals. Although overshoots are prevented at the receiver's input, reflections will still exist on the trace, as diodes do not affect trace impedance. To gain the benefits of both techniques, diodes may be used in conjunction with the other methods discussed herein to minimize reflection problems.

Termination not only matches trace impedance and removes ringing but will sometimes slow down the edge rate of the clock signal if the parameters are not set properly. Inappropriate termination could degrade signal amplitude and integrity to the point of nonfunctionality. Reducing either dI/dt or dV/dt will reduce RF emissions generated by high-amplitude voltage and current levels. Locate this resistive and/or optional capacitive (RC) network used for wave shaping immediately adjacent to the output pin of the clock driver. Locate terminations at the load end immediately at the end of the trace.

Another way to describe this dI/dt and dV/dt concern is to relate these functions to Ohms law, V = IR. The following text demonstrates how to translate Ohm's law into "simple" electromagnetic concepts.

If the impedance (Z) of the trace (resistor, inductor, bead-on-lead, etc.) is increased, then both dV (RF voltage) and dI (RF current) will decrease with the time variant pulse of the signal. With less RF voltage and RF current, less radiated and/or conductive RF energy will be generated, along with all EMI-undesirable side effects; hence, improved EMI performance will result. In addition to less RF currents, the edge rate of the signal may also be increased, along with a reduction of spectral RF energy. However, if the value of Z is too large, then nonfunctionality may occur due to excessive signal degradation. To guarantee proper functionality at all times, Z must be optimally calculated.

Calculate the series resistor R to be greater than or equal to the source impedance of the driving component, and lower than or equal to the line impedance, Z. This value is typically between 15 and 75 Ω (usually 33 Ω). The R value is lower if the end of the transmission line is terminated.

If an electrically long line must exist, this trace *must be terminated!* Long lines generally require use of high-current driving components. Calculate the terminating resistor value at the Thevenin equivalent of 50 Ω or the characteristic impedance of the trace. Use of "T-stubs" is not generally allowed. If a T-stub has to be used, the maximum permissible stub length cannot exceed $T = L_d^{Tr/10}$, and the lengths of each "T" arm from the center leg must be identical.

If a T-stub is required because of problems with layout or routing, it must be as short as possible. Use the measurement feature of the CAD system to measure routing lengths. If necessary, serpentine route the shorter trace until it equals its counter trace length exactly.

A potential or fatal drawback of using T-stubs lies in future changes to the artwork. If a different design engineer or CAD person makes change to the layout or routing to implement rework or a redesign, knowledge of this T-stub implementation may not be known, and accidental changes to the trace may occur, posing potential EMI and/or functionality problems.

4.14 CALCULATING DECOUPLING CAPACITOR VALUES

Capacitors can also be used to remove differential-mode RF currents on individual traces. Use of these parts is generally observed in I/O circuits and connectors and rarely in clock networks. The decoupling capacitor, C, alters the signal edge of the output clock line (slew rate) by rounding down the time period that the signal edge takes to transition from logic state 0 to logic state 1. This is illustrated in Fig. 4.13.

In examining Fig. 4.13, observe the change in the slew rate (clock edge) of the desired signal. The transition points remain unchanged; however, the time period t_r is different. This elongation or slowing down of the signal edge is a result of the capacitor charging and discharging as described by the equations detailed in Fig. 4.14. Note that a Thevenin equivalent circuit is shown without the load. The source voltage, V_b, and series impedance are internal to the IC or clock generation circuit. The capacitive effect on the trace, also seen in the figure, is a result of this capacitor being in the circuit. To determine the time rate of change of the capacitor in Fig. 4.13, use the equations in Fig. 4.14.

When a Fourier analysis is performed on this signal edge (conversion from time domain to frequency domain), we observe a significant reduction of RF energy plus a decrease in spectral RF distribution; hence, improved EMI compliance. Mathematical analysis of Fourier transforms is beyond the scope of this book. Excellent Fourier analysis is provided in many of the sources listed in the Bibliography.

There are two ways to calculate the capacitive value for decoupling. Remember that although capacitance is calculated for optimal filtering at a particular resonant frequency, use and implementation depends on installation, lead length, and other parasitic parameters that may change the resonant frequency of the capacitor. The capacitive reactance is the item of interest. Calculating the value of capacitance will get us into the ballpark, and it is generally accurate enough for actual implementation.

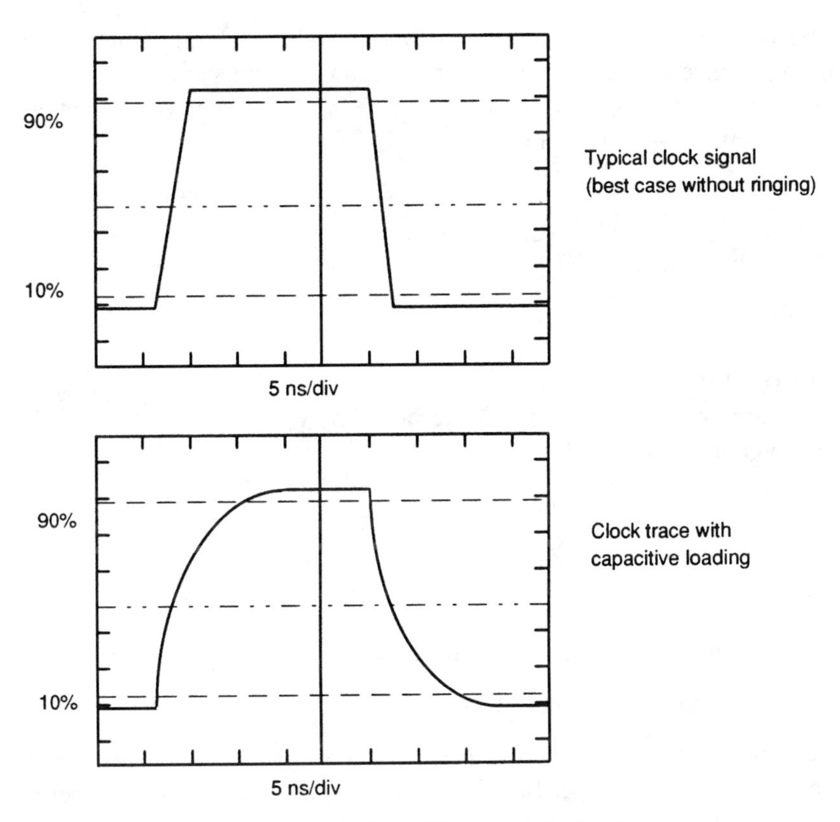

Fig. 4.13 Capacitive effects on clock signals

V b
R
Signal load
C
Capacitor Energizing

R
Signal load
C
Capacitor Discharging

Energizing

$$V_c(t) = V_b\left(1 - e^{-t/RC}\right)$$

$$I(t) = \left(\frac{V_b}{R}\right)e^{-t/RC}$$

Discharging

$$V_c(t) = V_o e^{-t/RC}$$

$$I(t) = \left(\frac{-V_o}{R}\right)e^{-t/RC}$$

Fig. 4.14 Capacitor equations

Before calculating a decoupling capacitor value, determine the Thevenin impedance of the network. This impedance value should be equal to the two resistor values placed in parallel. Using a Thevenin equivalent circuit, assume $Z_S = 150\ \Omega$ and $Z_L = 2.0\ k\Omega$

$$R_t = \frac{R_s R_L}{R_s + R_L} = \frac{150 \times 2000}{2150} = 140\ \Omega \qquad (4.21)$$

Method 1

To determine the capacitance value, and knowing the edge rate of the clock signal, use Eq. (4.22).

$$C_{max} = \frac{0.3 t_r}{R_t} \quad \text{or} \quad t_r = 3.3 \times R_t \times C_{max} \qquad (4.22)$$

where

C = nanofarads if t_r is in nanoseconds

C = picofarads if t_r is in picoseconds

The capacitor must be chosen so that $t_r = 3.2 \times R \times C$ equals an acceptable rise or fall time for proper functionality of the signal; otherwise, baseline shift may occur.

For example, if the edge rate is 5 ns, and the impedance of the circuit is 140 Ω, determine C to be

$$C_{max} = \frac{0.3 \times 5}{140} = 0.01\ \text{nF or } 10\ \text{pF} \qquad (4.23)$$

A 60 MHz clock with a period of 8.33 ns on and 8.33 ns off, R = 33 Ω (typical for an unterminated ALS part), has an acceptable t_r and t_f = 2 ns (25 percent of the on or off value). Therefore,

$$C = \frac{0.3\,(2 \times 10^9)}{33} = 20\ \text{pF} \qquad C = \frac{0.3 \times t_r}{R_t} \qquad (4.24)$$

Method 2

- Determine highest frequency to be filtered, f_{max}.
- For differential pair traces, determine the maximum tolerable value of each capacitor. To minimize signal distortion, use the Eq. (4.25).

$$\frac{1}{2\pi f_{max} \times \frac{C}{2}} \geq 3 \times R_t$$

$$C_{max} = \frac{100}{f_{max} \times R_t} \tag{4.25}$$

where C is in nanofarads and f in MHz.

To filter a 20 MHz signal with $R_L = 140\ \Omega$, the capacitance value would be

$$C_{max} = \frac{100}{20 \times 140} = 0.035 \text{ nF or } 35 \text{ pF} \tag{4.26}$$

When using decoupling capacitors, implement the following:

- If degradation of the edge rate is acceptable (generally 3 times the value of C), increase the capacitance value to the next highest standard value.
- Select a capacitor with proper voltage rating and dielectric constant for the intended use.
- Select a capacitor with a tight tolerance level. A tolerance level of +80/–0 percent is acceptable for power supply filtering but is ineffective as a decoupling capacitor for high-speed signals.
- Install the capacitor with minimal lead lengths.
- Verify that the functionality of the circuit still works with the capacitor installed. Too large a value capacitor can cause signal degradation.

4.15 COMPONENTS

Use frequency-generating components for system timing purposes only. Terminate unused drivers, buffers, and gates, or disable the output transistors by tying the input gate to logic "0" or "1" (ground or power, depending on the device), or disable the output enable pin for tri-state drivers.

Use oscillator modules for frequencies greater than 5 MHz or for clock edges faster than 5 ns When using oscillators 50 MHz and above, it is important that provision be made to add additional ground connections from the case of the oscillator to the ground planes through a wire, strap, or spring clip. The oscillator ground pin may be unreliable for grounding internally generated RF currents to ground; however, its use as a dc voltage reference pin is assured. Also, use a localized ground plane in clock-generation circuits as described in the beginning of this chapter.

Sharp turns (90° or more) affect the impedance of a routed trace and may create radiated RF currents as described in detail in Chapter 8. To prevent interaction between traces, employ a guard trace on one- and two-layer PCBs. This guard trace is connected by vias to the ground trace or planes at both the source and destination, in addition to random ground points along the periphery of the trace. The main function of the guard trace, when used in this application is:

- to prevent crosstalk from high-threat signal traces to nearby components or other traces, thereby maintaining signal quality
- to provide a controlled return path for RF currents to their source
- to provide a minimum signal-to-return loop area

4.16 TRACE SEPARATION AND THE 3-W RULE

Crosstalk may exist between traces on a PCB. This undesirable effect is not only associated to clock and periodic signals but to other system-critical nets. Data, address, control, and I/O traces all are affected by crosstalk and coupling. Clocks and periodic signals present the majority of problems in a PCB and can cause functionality problems with other functional subsections. Use of the 3-W rule will allow a designer to comply with PCB design criteria without having to implement guard traces.* Note that the "3-W rule" represents the approximate 70 percent flux boundary at logic currents. For the approximate 98 percent boundary, use "10-W."

The basis for use of the 3-W rule is to minimize coupling between traces and signals and to provide a "clean path" along the board where the signal flux and the return flux will link and cancel properly, without the perturbations of vias or flux from other traces. The 3-W rule states that *"the distance separation between traces must be three times the width of the traces, measured from centerline to centerline,"* or otherwise stated, "the distance separation between two traces must be greater than two times the width of a single trace." For example, assume that a clock line is 6 mils wide. No other trace can exist within a minimum of 2 × 6 mils of this trace, or 12 mils, edge-to-edge. As observed, much real-estate is lost in areas where trace isolation occurs. An example of the 3-W rule is shown in Fig. 4.15.

* The terms *3-W rule* and *10-W rule* were first described and defined by W. Michael King.

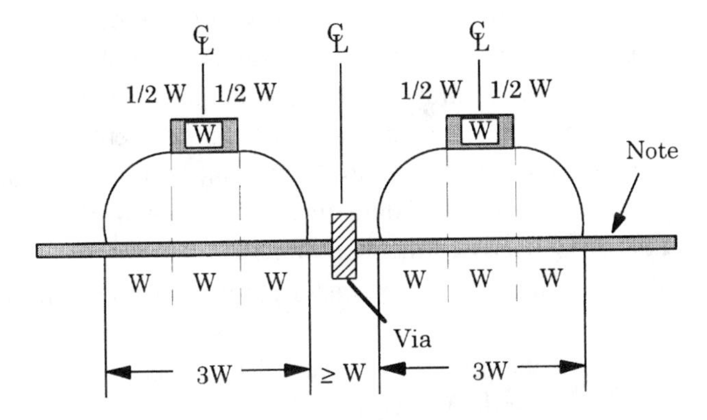

For the via, add annular diameter which includes both the via and annular clearance

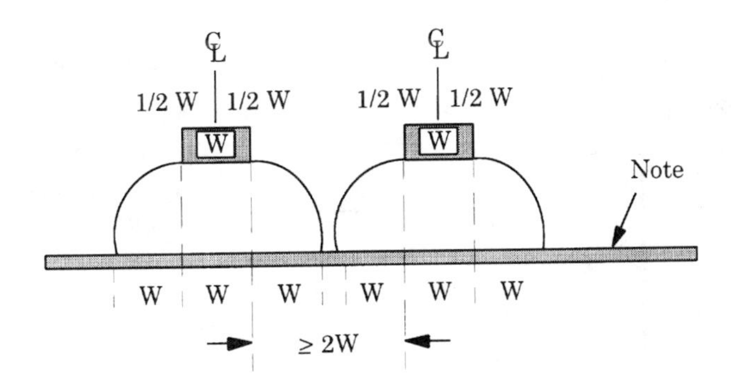

The distance spacing between both traces must have a mimimum overlap of 2W

Note: **Application of the 3-W rule to traces routed along board edge requires ≥ 1-W from the outside edge of the trace to the edge of the ground plane.**

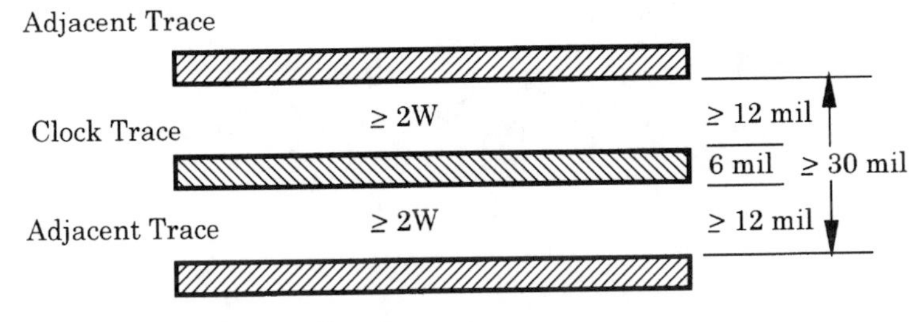

Top down view
3W spacing without a via between the traces

Fig. 4.15 Designing with the 3-W rule

Do not restrict use of the 3-W rule to only clock traces. Differential pairs (balanced, ECL, etc.) are prime candidates for 3-W. Power plane noise will capacitively (or inductively) couple into the paired traces causing data corruption. In I/O sections where differential traces are routed with an absence of copper on adjacent planes (no image plane), an alternative technique used is to route these traces using the 3-W rule. (Absence of copper is discussed in Chapter 5.)

Differential pair traces, if not routed in parallel on the same routing plane (due to lack of available board space), must be routed on adjacent planes (absence of copper area). One trace must physically be three times the width of its corresponding trace along the entire route, as shown in Fig. 4.16. This technique helps minimize RF fringing between traces.

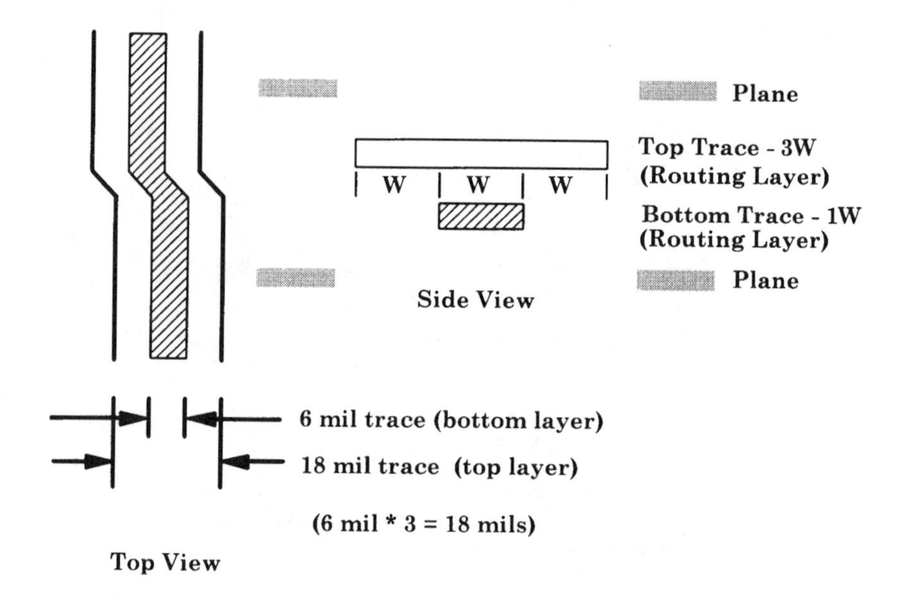

Fig. 4.16a Differential pair routing and the 3-W rule (vertical axis)

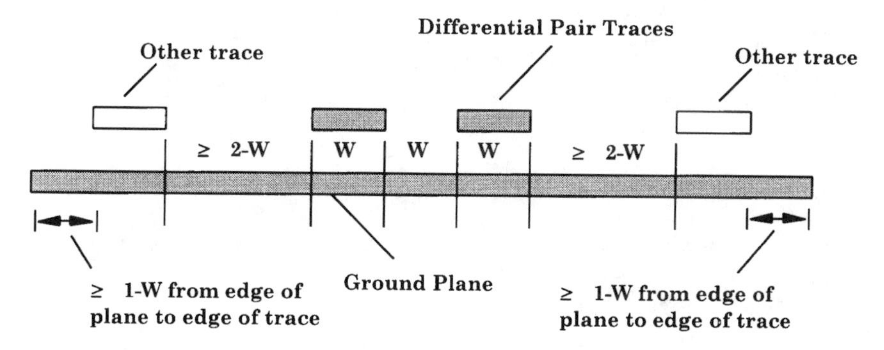

Fig. 4.17b Parallel differential pair routing and the 3-W rule (horizontal axis)

4.17 REFERENCES

1. Mardiguian, M. 1992. *Controlling Radiated Emissions by Design.* New York: Van Nostrand Reinhold.
2. Motorola, Inc. *MECL System Design Handbook* (#HB205) and *Transmission Line Effects in PCB Applications* (#AN1051).
3. Motorola, Inc. *Transmission Line Effects in PCB Applications* (#AN1051/D).
4. Dockey, R.W, and R.F. German. 1993. New techniques for reducing PCB common-mode radiation. *Proceedings of the IEEE International Symposium on Electromagnetic Compatibility.* New York: IEEE, 334–339.

5

Interconnects and I/O

One of the more sensitive parts of a PCB, in terms of RFI, ESD, and other forms of radiated and conducted susceptibility, are the I/O and related interconnects. These include front panel display indicators and controls, serial and parallel ports, Ethernet connectors, external SCSI interconnects, modems, video and audio cables, power cords, blank adapter brackets, peripheral drive cover plates, input devices (mice, keyboard, joysticks, hand scanners), transducers, network cabling, and a range of other peripherals.

Most EMI emissions in I/O circuitry are generated from a combination of:

- common-mode coupling inside I/O interface devices
- power plane noise coupled into I/O circuits and cables
- clock signals coupling onto I/O cables through both conductive and radiated modes
- RF energy coupling onto cables exiting the enclosure
- lack of data line filtering on connectors and signals traces (common-mode and differential-mode)
- improper connection of chassis, signal, frame, digital, and analog grounds
- use of various I/O connectors (plastic versus metal or unshielded versus shielded)

I/O circuitry may generate as many problems with electromagnetic interference and susceptibility concerns as do clock signals—perhaps more. Proper component selection and placement minimizes RF coupling that may occur with both conducted and radiated emissions. I/O must also be physically (electrically) separated on the PCB from high RF bandwidth components. If possible, separation from medium RF bandwidth circuits is also recommended, depending on the application.

One example of proper I/O implementation is to have metal I/O connectors RF bonded to chassis ground via a low-impedance path (discussed in detail later in this chapter) to facilitate the correct bonding of cable shielding to the casework. This low-impedance path must consist of a 360° solid bonding of the metal connector housing to chassis ground. In addition, the designer should provide direct grounding of signal and/or chassis shield grounds immediately at the connector entrance point *without use of a pigtail* for circuits operating at 1 MHz and above.

I/O drivers and receivers must be physically located as close to the I/O connector as possible to minimize trace lengths and to minimize the risk that these signals will receive coupling from other signals. Filtering of data signals is often required. This filter is placed between the driver/receiver devices and connectors.

5.1 PARTITIONING

Partitioning I/O circuits involves three basic areas of concern. These are functional subsystems, quiet areas, and radiated noise coupling. Each is briefly discussed below, with more details presented later.

5.1.1 Functional subsystems

Each I/O should be considered as a different subsection on a PCB, as each may be unique in its particular application. To prevent RF coupling between subsystems, isolation is required. A functional subsystem consists of a group of components and their respective support circuitry. Locating components close to each other minimizes trace length routing and optimizes functional performance. Every hardware design and CAD engineer generally tries to group components together; however, for various reasons, it is sometimes impractical to do so. I/O subsystems must still be treated differently during layout from any other section of the PCB. This is generally done through layout partitioning, as discussed in greater detail later.

Layout partitioning enhances signal quality and functional integrity by preventing high-bandwidth RF signal emitters (e.g., backplane interconnects, video devices, data interfaces, Ethernet controllers, SCSI devices, CPUs) from corrupting serial, parallel, video, audio and asynchronous/

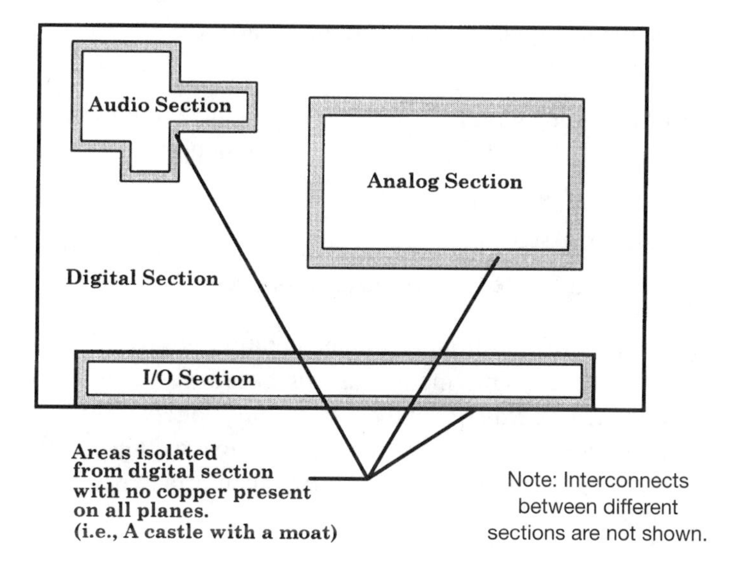

Fig. 5.1 Quiet areas

synchronous ports; floppy controllers; front panel console displays; local and wide area networks controllers, and so on. Each I/O subsystem must be conceived, designed, and treated as if the subsystems were separate PCBs.

5.1.2 Quiet areas

Quiet areas are sections that are physically isolated from digital circuitry, analog circuitry, and power and ground planes. This isolation prevents noise sources located elsewhere on the PCB from corrupting susceptible circuits. An example is power plane noise from digital circuits entering the power pins of analog devices, audio components, I/O filters and interconnects, and so on, as shown in Fig. 5.1.

Each and every I/O port (or section) must have a partitioned (quiet) ground/power plane. Lower-frequency I/O ports may be bypassed with high-frequency capacitors (usually 470 pF to 1,000 pF) located near the connectors.

Trace routing on the PCB still must be controlled to avoid recoupling RF currents into the cable shield. A clean (quiet) ground must be located at the point where cables leave the system. Both power and ground planes must be treated equally, as these planes act as a path for RF return currents. RF return currents from components other than I/O can inject switching high bandwidth RF noise into the I/O area and components.

To implement a quiet ground, use of a partition or moat is required. This quiet area may be:

1. 100% isolated with I/O signals entering and exiting via an isolation transformer (i.e., Ethernet AUI type) or an optical device;
2. data line filtered; or
3. filtered through a high-impedance common-mode inductor or protected by a ferrite bead-on-lead component.

Another method of partitioning implementation is possible with any of these three methods. This method is composed of a single entry point between *dirty power plane(s)* and *clean or quiet zone(s)*. This single entry point is called a *bridge,* as described in the next section. (If the "quiet" zone is a "castle" surrounded by a "moat," then it may be entered with a "bridge" across the moat.)

5.1.3 Internal radiated noise coupling

Radiated RF coupling can occur between different functional subsections. To prevent internal RF coupling (e.g., internal disk drive cable signal noise to I/O connectors, radiated RF currents from CPU components to other components, I/O controller logic to I/O cables), a *fence* may be required. A fence is a metal barrier secured to the ground plane(s) at intervals appropriate for the highest frequencies ($\lambda/20$ wavelength) anticipated and tall enough in height to prevent direct line RF radiated coupling between components. A fence is similar to one side panel of a metal case that would normally encapsulate a circuit or device, except it is mounted on the circuit board. This fence is also similar to the standard bus bar used for power and ground distribution on the top side of a PCB. At each and every ground post location, decouple absorbed RF currents using bypass capacitors between the grounded fence and the power and ground planes around the fence.

Determine in advance if logic circuitry or subsections are candidates for emissions of, and susceptibility to, internally generated RF currents. Depending on placement of components on the PCB, relative to susceptible circuits or I/O connectors, anticipate potential coupling of internal RF energy before routing traces or finalizing placement. Design provisions for possible future installation of a fence may save a design, since a fence is easily incorporated if mounting vias are provided. Actual implementation and use will be determined through functional testing. It is easier to add a fence to a PCB (addition to the bill of material) than to incorporate this fence in the artwork after the board has been manufactured.

5.2 ISOLATION AND PARTITIONING (MOATING)

Isolation and partitioning refers to the physical separation of components, circuits, and power planes from other functional devices, areas, and subsystems. Allowing RF currents to propagate to different parts of the board by radiated or conductive means can cause problems not only in terms of EMI regulatory compliance but also with regard to functionality.

Isolation is created by an absence of copper on *ALL* planes of the board. Absence of copper is created using a wide separation (typically 50 mils minimum) from one section to another. In other words, an isolated area is an island in the board, similar to a castle with a moat. Only those traces required for operation or interconnect can travel to this separate area. The moat serves as a "keep out" zone for signals and traces that are unrelated to the moated area or its interface. Two methods exist to connect traces, power, and ground planes to this island, as described in the following text. Method 1 uses isolation transformers or optical isolators and common-mode data line filters to cross the moat. Method 2 uses a bridge in the moat. Isolation is also used to separate high-frequency-bandwidth components from lower-bandwidth circuits while maintaining low-EMI bandwidth I/O in terms of the RF spectrum.

5.2.1 Method 1: Isolation in moating

Method 1 involves the use of an isolation transformer or optical isolators. An I/O area is sometimes 100 percent isolated from the rest of the PCB. Only at the metal I/O connector is RF bonding to chassis ground performed, and then only through a low-impedance path to ground *outside this isolated area.* Use of bypass capacitors from shield ground (or braid) of the I/O cable to chassis ground is sometimes needed in place of a direct connection when required by the interface specification. *Shield ground* (or *drain wire*) refers to a discrete pin or wire in the connector that connects the internal drain wire of the external I/O cable to its mylar foil shield, also internal to the cable. Some engineers tie shield ground to chassis ground through a pigtail. *Pigtails are acceptable only for audio circuits and not for high-frequency signals or components. Pigtails are also susceptible to external high-frequency EMI threats and audio rectification, which may cause system-wide failures.*

There are two areas of concern for selection of bypass capacitors used in I/O circuits:

1. proper bandwidth filtering, and
2. peak surge voltage protection capabilities for electrostatic discharge protection

For example, Ethernet circuits require use of an isolation transformer for compliance with ISO/IEC 8802.3 to physically isolate the network from the system (computer) in case an abnormal fault occurs in the controller, thus maintaining network quality. Common-mode data line filters may be used in conjunction with isolation transformers. Common-mode data line filters (usually toroidal in construction) may be used in both analog and digital applications. These filters remove common-mode RF currents carried on signal traces that would exit through the I/O cable. If power and ground are required in the isolated area (i.e., +12 Vdc for Attachment Unit Interface [AUI]), cross the moat with a ferrite bead-on-lead for the power trace and a single solid trace for ground. Locate the secondary short-circuit fuse (required for product safety) on either side of the ferrite bead. Sometimes capacitive decoupling is required to remove digital noise from I/O power. Locate the optional decoupling capacitor, with one terminal of the capacitor to the filtered side of the ferrite bead (output side) and the other terminal to the digital ground plane. Locate the power filtering components across the moat at the far outside edge of the board. Route both power and ground trace adjacent to each other to minimize the RF ground loops that can be developed between these two traces if located on opposite sides of the moat. This arrangement is shown in Fig. 5.2.

5.2.2 Method 2: Bridge in a moat—partitioning

Method 2 uses a bridge between a control section and a partitioned area. A bridge is a break in the moat, at only one location, where signal traces, power, and ground cross the moat. This is illustrated in Fig. 5.3. Violation of the moat by any trace not associated with the I/O circuit can cause problems. RF loop currents will be created as detailed in Fig. 5.4. RF currents must image back along their trace route, or common-mode noise will be produced between the two separated areas. Unlike Method 1, power and ground planes are directly connected between the two areas, hence this method forms a partition.

Sometimes, only the power plane must be isolated and the ground plane connected at the bridge. This technique is common for circuits where a common ground plane is required, or separately filtered or regulated power is required. In this case, a ferrite bead-on-lead or inductor is required to bridge the moat for the filtered power only. Locate this bead in the bridge area and not over the moat. If analog or digital power is not required in the isolated area, redefine this now unused power plane as a second ground plane referenced to the main ground plane by vias.

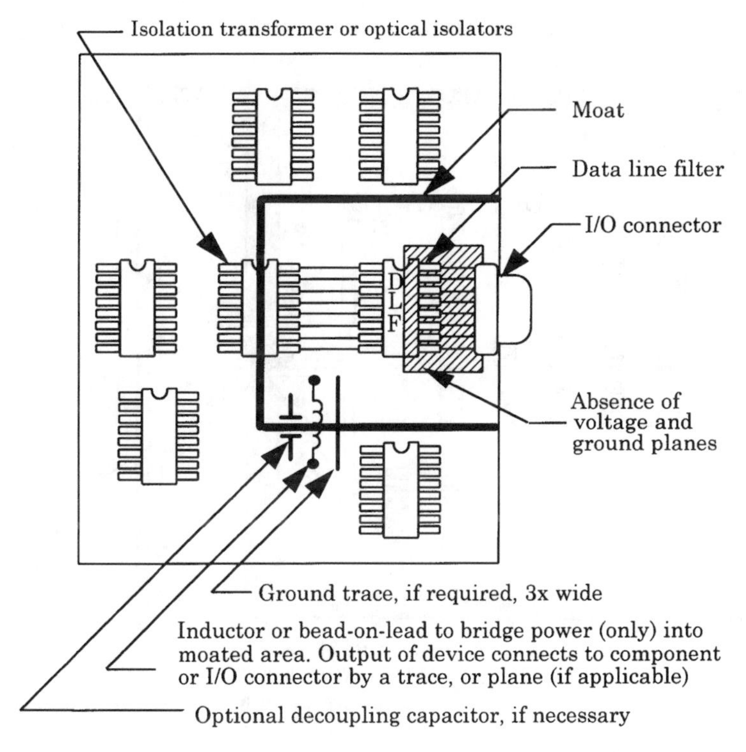

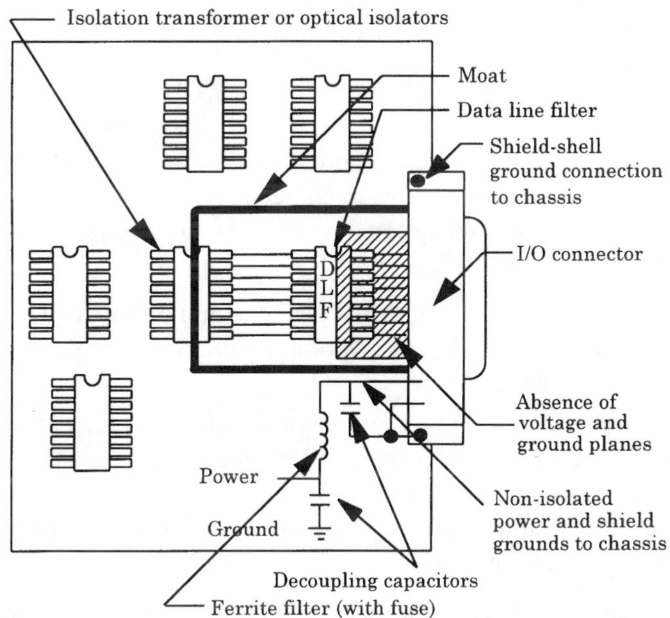

Fig. 5.2 Method 1: Isolation in moating, with (top) isolated and (bottom) non-isolated AUI port power

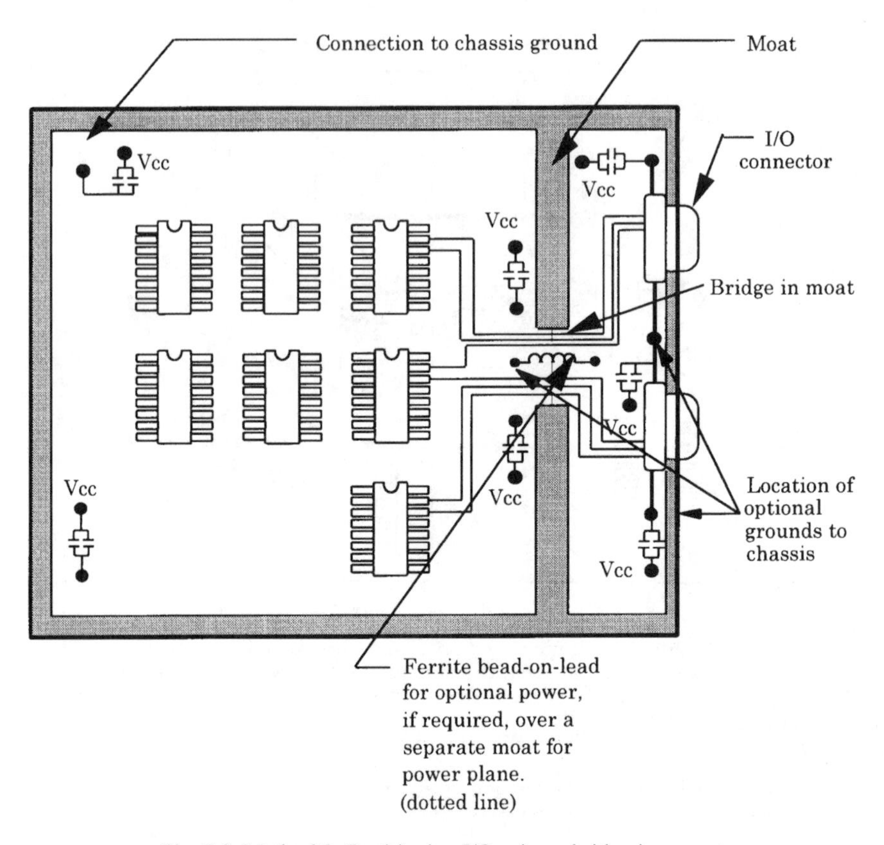

Fig. 5.3 Method 2: Partitioning I/O using a bridge in a moat

When using bridging method 2, grounding both ends of the bridge to chassis or frame ground is highly recommended if multipoint grounding is provided in the chassis and system-level design. Grounding the entrances to the bridge performs two functions:

1. Grounding prevents high-frequency common-mode RF components in the ground planes (ground-noise voltage) from coupling into the partitioned area.
2. Grounding helps remove eddy currents (for improved ground loop control) that may be present in the chassis or card cage. A much lower impedance path to chassis ground is provided for RF currents that would otherwise find their way to chassis ground through other paths, such as RF currents in an I/O cable.

Grounding both ends of the bridge also increases electrostatic discharge immunity. If a high-energy pulse is injected into the I/O connector, this energy may travel to the main control region and cause permanent

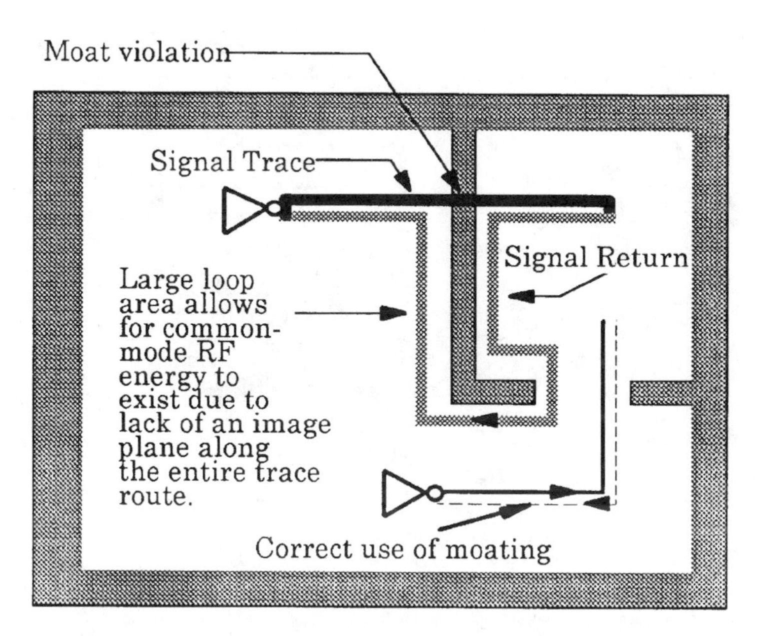

Fig. 5.4 Violation of the moating concept (Source: CKC Laboratories, reprinted by permission)

damage. This energy pulse should be sunk to chassis ground through a very low-impedance path.

Another reason to ground both sides of a bridge is to remove RF ground-noise voltage created by voltage gradients that appear between the partitioned area and main control section. If the RF common-mode noise contains high-frequency RF components, decoupling capacitors for voltages should be provided at each chassis ground stitch connection.

Figure 5.5 illustrates how traces are to be routed when using both digital and analog partitions. Since power plane switching noise may be injected into the analog section from digital components, isolation and/or filtering is required—especially on the power plane. All traces that travel from the digital to analog sections must be routed through this bridge. For analog power, use an inductor or ferrite bead-on-lead to cross the moat. A voltage regulator may also be required. The analog power moat is usually 100 percent complete around the entire partition. Certain analog components want analog ground to be referenced to digital ground, but only through this bridge as shown in Fig. 5.3. Many analog-to-digital and digital-to-analog devices connect the "AGNDS" and "DGNDS" (indicated on the pin designation) together in the device lead frame. When such is the application of a partition, one ground, bridged together, is required or the digital signal currents will not return efficiently to their source, causing

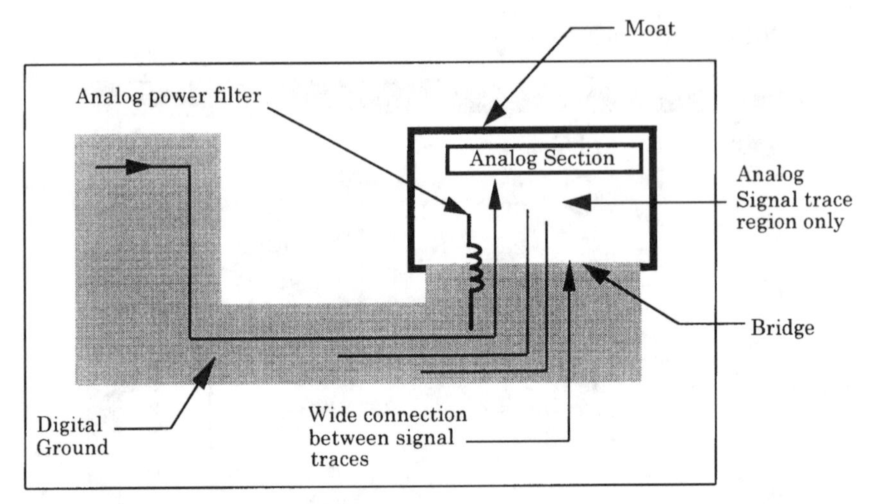

Note: All signal traces must pass through this region only (bridge).
No signals are to pass over a plane void region (moat).
For the analog section, the power plane is 100% moated.
If a bridge is used for ground, both digital and analog ground will be at the same potential.

Fig. 5.5 Digital and analog partitioning

noise and EMI. AGNDS and DGNDS should be moated away from each other only when the circuit devices themselves provide the AGND to DGND isolation.

5.3 FILTERING AND GROUNDING

5.3.1 Filtering

There are two basic types of RF filters: capacitive and inductive. Different applications require one, the other, or both. Most lower-bandwidth (≤ 1 MHz) interfaces will benefit from the use of filtering at the ports. These filter components will not be effective unless their placement is exactly adjacent to their entry point. One inch (2.54 cm) is too far away. Capacitive bypass filtering is used to remove high-frequency RF currents from external I/O cables as well as differential-mode RF current from logic devices and I/O interconnects.

For both radiated emissions and radiated susceptibility, locate any bypass capacitors directly at the connector entry point. Position the data line filter (inductor or common-mode choke) between the controller side of the signal trace and the I/O connector with a bypass capacitor between, shown in Fig. 5.6, Technique #1. These filter capacitors are especially critical with keyboard and mouse I/O interconnects and external cables that may not be shielded, or have minimal shielding.

For both radiated emissions and electrostatic discharge events, a different arrangement is required. Locate the bypass capacitor at the input side of the data line filter and not at the I/O connector as seen in Fig. 5.6, Technique #2.

The reason why these two design techniques are application dependent will become evident in the individual discussions that follow.

Technique #1: Bypass capacitors at the I/O connector

Figure 5.6 shows filter capacitors located directly at the I/O connector. These capacitors are excellent filter elements for RF currents emanating from the product (radiated and/or conducted emissions). Selection of a capacitor with a very high self-resonant frequency is required for optimal performance while minimizing degradation of the signal edge required for proper operation of the I/O signal. Values of these capacitors typically range between 1,000 pF and 100 pF, depending on the signal bandwidth.

In the case of an ESD event, voltage and current levels may be extremely large in amplitude. International test specifications mandate a minimum voltage level of 1500 V, with 6000 and 8000 V also used as a common reference value. Surface mount capacitors used for RF emissions are usually rated 25 V or less. Should an ESD event be injected into the I/O line, these filter capacitors could self-destruct as the damage level (volt) is generally much higher. Capacitor component failure may occur as a function of overvoltage and/or overstress as an artifact of the excessive voltage/current through the component body. Capacitor component rating should be evaluated considering the actual ESD voltage that will be developed across the bypass capacitor (which is usually less than the ESD source voltage because of ESD loading by the bypass capacitor) and the peak current impulse stress through the capacitor. When this condition occurs, capacitive filtering for the I/O lines no longer exists. This leaves the unit noncompliant for both RF emissions and immunity purposes. For this reason, caution is urged in component selection.

Technique #2: Bypass capacitors at data line filter input

International EMC requirements mandate that products must not only comply with radiated emission limits but must provide a level of immunity to externally induced RF currents. These RF currents may be a result of an ESD event, a fast transient burst or pulse, radiated or conducted fields

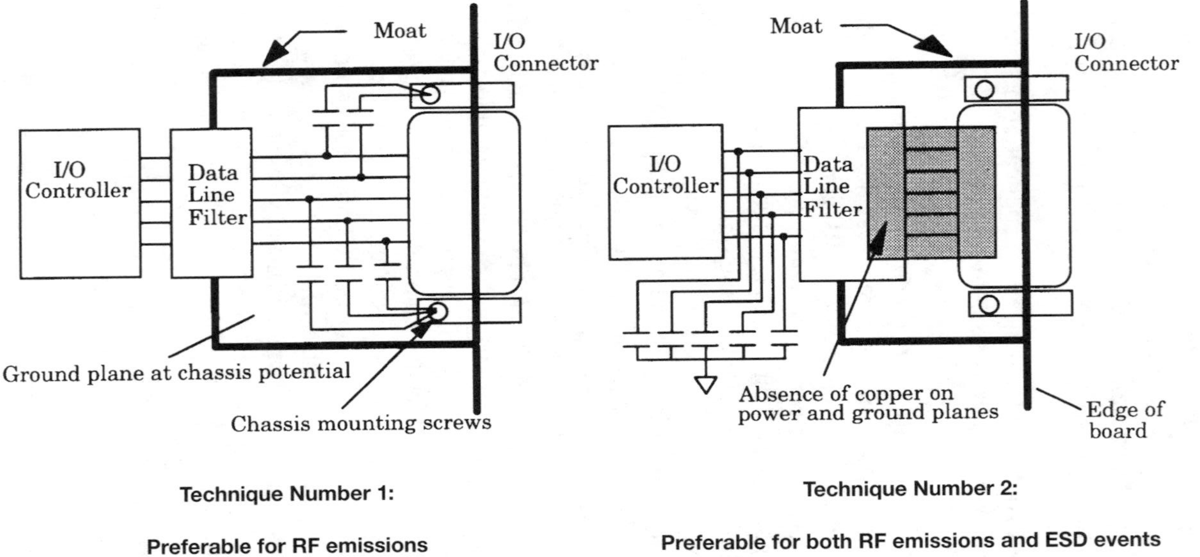

Notes:

Moat travels directly through the center of the data line filter.

Depending on I/O port and connector type, all or some of the connector may be within the moat.

Capacitor (filter) values depend on signal bandwidth.

Metal shell of connector is connected to chassis ground, not logic or signal ground.

Fig. 5.6 I/O filtering

from adjacent products, man-made RF energy (cellular telephones, pagers), and environmental disturbances.

To minimize stress to low-voltage-rated surface mount capacitors and still receive the benefit of a bypass capacitor, locate the bypass capacitor on the input side of the data line filter. This is shown in Fig. 5.6, Technique #2. When an ESD event occurs, the pulse will first see a high-impedance source—the data line filter. The filter will prevent the full magnitude of the event from reaching control circuitry or other logic devices on the main part of the PCB. ESD is partially captured between the data line filter and I/O connector. If a circuit is sensitive to a particular frequency, a resonant-shunt LC filter could be designed to alleviate this problem by installing the capacitor. Provisions should always be made for use of bypass capacitors during layout in applications where it is known that the capacitors will not impact functionality. Actual use eventually will be determined through functional testing.

To calculate the value of this bypass capacitor, use Eq. (5.1). If an inductor is used instead of a data line filter, calculate the value of the capacitor, since the value of L is known indirectly.

$$f_r = \frac{1}{2\pi\sqrt{LC}}$$

$$C = \frac{\left(\frac{1}{2\pi f_r}\right)^2}{L} \tag{5.1}$$

where

f_r = resonant frequency

L = inductance of the circuit (including L of the capacitor leads)

C = capacitance value

When using a data line filter, the inductance value, L, is replaced by inductive reactance. Because of this LC combination, design a filter for a null resonance (if applicable) at the particular frequency of concern. For this to occur, inductive reactance must equal capacitive reactance, $X_L = X_c$, where

$$X_L = 2\pi fL$$

$$X_c = \frac{1}{2\pi fC} \tag{5.2}$$

Design techniques for PCB layout have been presented for I/O interconnects. The question that remains to be answered is what the difference is between common-mode and differential-mode voltage and current as related to these circuits. Why are inductive components used for one application and capacitors for another? Common-mode current can be induced on cables attached to a PCB by differential-mode currents on the ground plane. Common-mode currents frequently are produced by sources that originate as "differential-mode to common-mode conversion." In execution, imbalance currents or non-cancelling differential-mode flux result in common-mode by-products of the sources simply because often they are not perfectly balanced or cancelling. This ground noise in the plane relates differential-mode voltage and currents on the board to common-mode voltage and current on I/O cables, and even across the board planes.

A decoupling capacitor (with an extremely small loop at point of use) provides a current-driven mechanism for driving a common-mode antenna. Large power switching cross-conduction device currents produced by components will produce a voltage drop on the ground trace or plane. This voltage drop can potentially drive two portions of the ground systems against one another (as two halves of an antenna). Common-mode voltage reflection in signals between sources and connectors may be increased by attempting to eliminate the voltage-driven mechanism by placing a decoupling capacitor between the power and ground source at the I/O connector or cable. What this means is that, although the cable port may be suppressed, the board itself may radiate more, or at different frequencies (because of L-C resonance between trace and path inductance and the bypass capacitors). The length of the differential-mode loop is increased by the addition of the decoupling capacitor at the connector physically located distant from the source component. An increase in loop inductance will be created (from source to capacitor); hence, an increase in voltage drop is observed along the return conductor (the ground plane) [1].

Inductive filtering is used in series with I/O signal lines to remove common-mode RF currents in the signal trace. This method of filtering is achieved using a ferrite bead or other form of inductance. Low-loss inductors make poor EMI filter elements. Use inductors that are RF energy absorbers.

A detailed discussion on how to select the proper ferrite device (bead-on-lead, chip bead, toroidal shield bead, toroidal core, etc.) for RF current suppression (common-mode) is presented in Chapter 8.

When data line filters (common-mode, common-coil, toroids) are used to bridge two areas, common-mode RF currents are removed from the signal trace. Selection of the correct material type of data line filter is important. This is because the ferrite material and core construction are designed for optimal performance within a specific range of frequencies (explained in Chapter 8). One manufacturer's data line filter or ferrite material may provide 45 dB attenuation at 30 MHz and 10 dB attenuation at 100 MHz, whereas another vendor's product (nearly identical in form and fit but with different characteristic profile permeability) may provide 15 dB attenuation at 30 MHz and 40 dB at 100 MHz. Verify that the permeability and composition of the ferrite material (in short, the total impedance) is compatible with the intended range of frequencies to be suppressed.

5.3.2 Grounding (I/O connector)

For products that are low-frequency and that may use *single-point grounding,* this section is generally not applicable. For such low-frequency products, low-impedance connections between logic ground and chassis ground can cause not only susceptibility to electromagnetic interference but may also prevent proper functionality in the form of signal-to-noise degradation. This is especially true for audio analog circuits that are devoid of digital processing. For a circuit to qualify as "low frequency" to the extent that it also qualifies for single-point grounding methods, it is required that the combination of signal levels, packaging techniques, and all operating frequencies be such that transfer currents to the case (or external surfaces) through distributive transfer impedances be insignificant in comparison to the operative signal levels *or* the desired EMC criteria!

For products using *multipoint grounding,* this section is applicable wherever an I/O interface is used. Most modular PCBs contain a mounting bracket, faceplate, bulkhead connector, or a securement means between control logic and the outside world. This securement may contain various I/O connectors, or it may be a blank panel (i.e., EISA/ISA adapter bracket). This bracket must be RF bonded by a low-impedance metal path directly to chassis ground. This bracket grounding may also be bonded to logic ground.

Provide for multiple ground connections from the ground planes of the PCB to the I/O bracket. Multiple ground points in the appropriate locations redirect RF ground loops between grounding locations on the bracket, distributive transfers to the case, and the opposite end of the PCB. The better the grounding, the more sourcing of RF currents to chassis ground. Figure 5.7 shows how to properly ground a mounting bracket to both chassis and logic ground. All I/O areas are isolated from control logic by a moat.

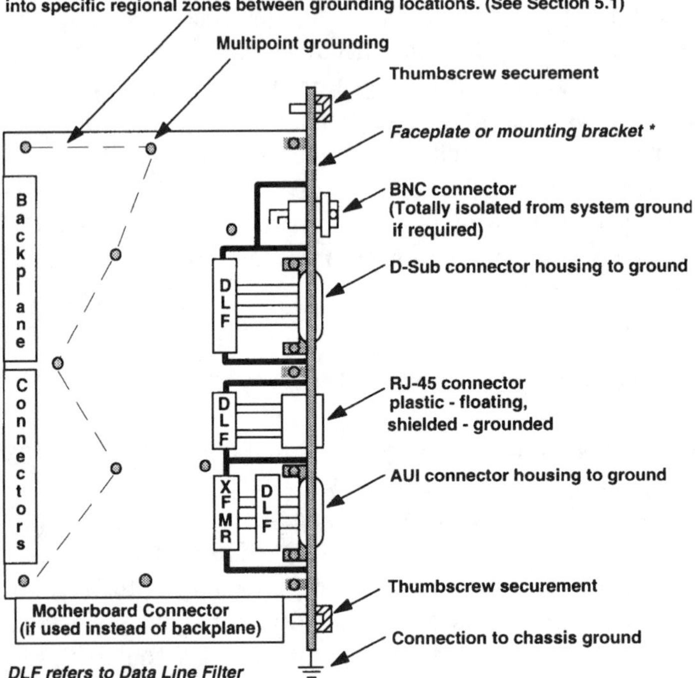

Fig. 5.7 Multipoint grounding of I/O faceplate or bracket (see also Fig. 5.6)

5.4 LOCAL AREA NETWORK I/O LAYOUT

Local area networks (LANs) and wide area networks (WANs) require careful attention during layout to ensure compliance with international protocol interface and emission requirements. LANs come in various topological (e.g., ring, star) and technological (e.g., broadband, baseband, token passing, carrier sense multiple access) configurations. Many variations are sold commercially, including the omnipresent Ethernet. A brief sampling of network specification documents includes:

ISO/IEC 8802–3
: *Carrier Sense Multiple Access with Collision Detection (CSMA/CD) Access Method and Physical Layer Specifications (Ethernet).*

ISO/IEC 803–4

Token Bus Access Method and Physical Layer Specification.

ISO/IEC 8802–5

Token Ring Access Method and Physical Layer Specifications. Subsections describe STP (Shielded Twisted Pair), UTP (Unshielded Twisted Pair) and Fiber Optic schemes.

ISO 9314–X

Fiber Data Distributed Interface (FDDI)—Token Ring Physical Layer Medium Dependent (PMD). (Consists of five subsections describing various details of this series.)

"Ethernet" has several different specifically defined interface formats, each with a different front-end connector and special design considerations. The most widely used Ethernet-type protocols are tabulated below. This list is not intended to be comprehensive; it is only to illustrate some of the variations within one type of LAN.

10Base-5	(10 MHz, AUI—Coax)
10Base-2	(10 MHz, Thinnet—Coax)
10Base-T	(10 MHz, RJ-45,—dual twisted pair, shielded or unshielded)
10Base-F	(10 MHz, FOIRL—Fiber Optic)
10Base-FL	(10 MHz, FL—Fiber Optic)
100Base-TX	(100 MHz, 100Base-T using 2 pairs of Category 5 UTP cable)
100Base-T4	(100 MHz, 100Base-T using 4 pairs of Category 3, 4 or 5 UTP cable)
100VG-AnyLAN	(100 MHz, Uses Category 3, 4, or 5 UTP, STP cable and Fiber Optic)

The recommended design implementations for some typical LANs are illustrated in Figures 5.8 to 5.11. Similarities exist between different media interfaces. Observe the layout concept and implementation. Specific design details are left to the designer and may differ based on specific implementation and the component placement required for each Ethernet and/or token ring application. The order of components from the LAN controller are isolation transformer/wave shaping circuit, data line filter, and I/O connector.

The recommended layout topologies shown in Figures 5.8 through 5.11 have been widely used to allow compliance with international Class B emission requirements. Proper design of the front-end interface is not the

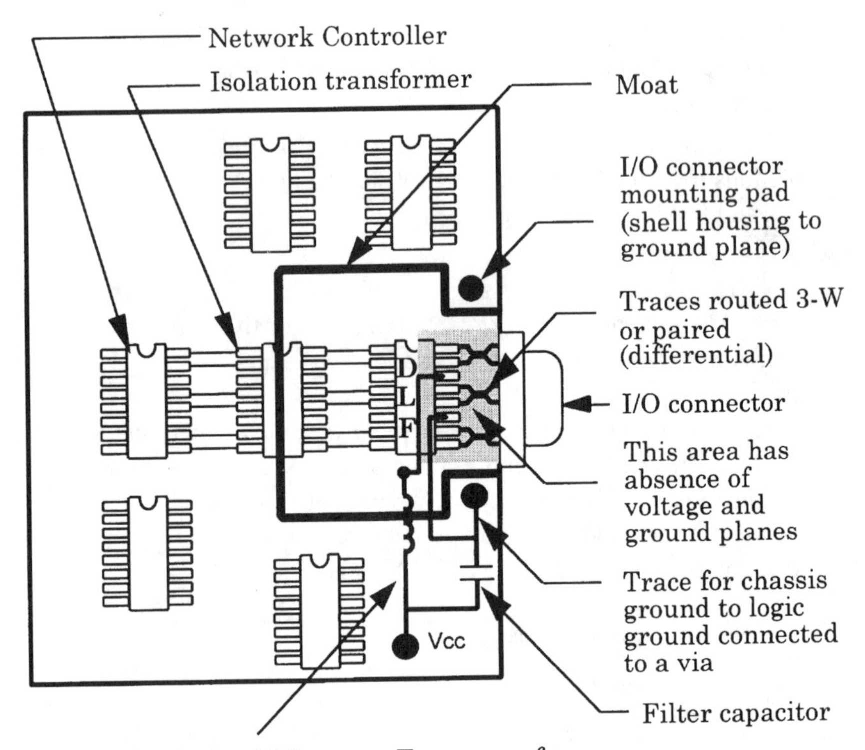

Fig. 5.8 Recommended layout for AUI, 10Base-T (RJ-45) and token ring (AUI uses three differential pairs, whereas 10Base-T and token ring use two pairs of differential traces.)

total solution for electromagnetic interference compliance. All other areas discussed in this guide are still required—particularly with regard to clock circuitry, trace routing, and correct I/O isolation and filtering.

An important design concern in the front end interface for Ethernet and token ring lies in input shunt capacitance. The ISO/IEC 8802–X standard specifies the maximum amount of shunt capacitance permitted between the waveshape/filter/isolation transformer module and the I/O cable. Use of a common-mode choke (data line filter) can add capacitance, which may degrade signal functionality/quality in addition to violating ISO/IEC specifications. If use of a common-mode choke is desired, select and test devices as they would be installed in the final layout for compliance with network interface specifications.

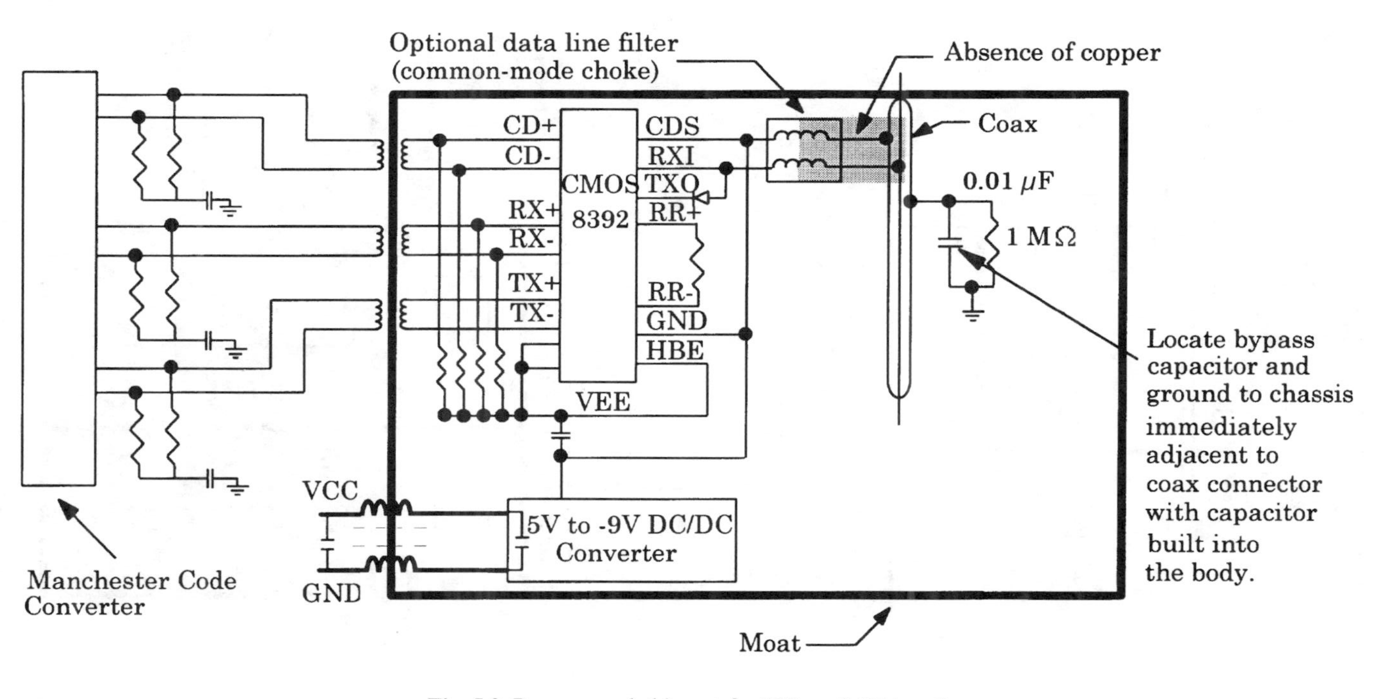

Fig. 5.9 Recommended layout for 10Base-2 (Thinnet)

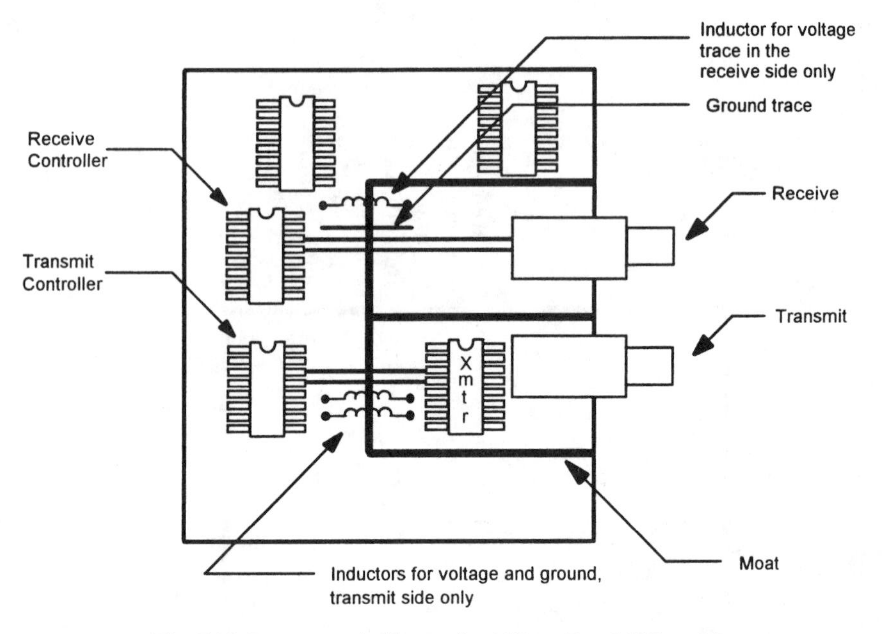

Fig. 5.10 Recommended layout for 10Base-F and 10Base-FL

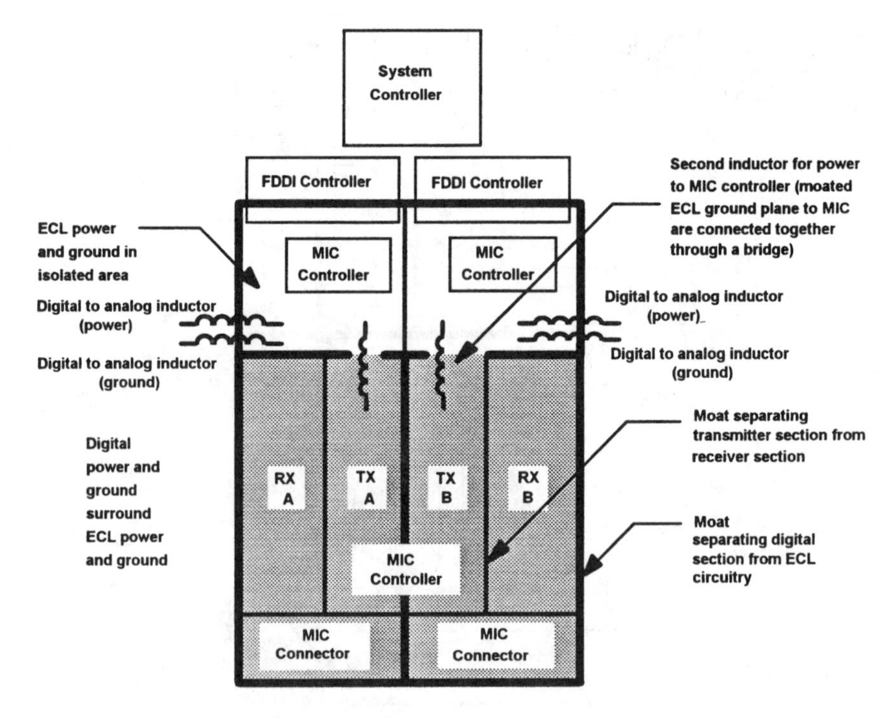

Fig. 5.11 Recommended layout for FDDI

5.5 VIDEO

PCBs with video interface require careful attention to impedance control, filtering, and grounding. For analog monitors, provide the slowest slew rate signal possible from the video generator to the monitor. Occasionally, the video interface signal can be substantially slowed down in edge times to minimize differential-mode to common-mode EMI conversion (and EMI emission) in the video cable, since the monitors may "re-square" the interface signals as preconditions for video drive within the monitor. Install this passive filter between the video generator and I/O connector, with the capacitor located immediately adjacent to the connector with minimal or zero trace lead length. Manufacturers of video controllers generally recommend a particular way to lay out the front end design, including selection of discrete components. For analog monitors, a constant trace impedance of the three RGB signals, along with both the horizontal and vertical sync traces, is required to prevent reflections and standing waves due to transmission line impedance mismatches in the system and interconnect cable. Often, substantial filtering may be applied to the vertical and horizontal sync signals without functional degradation since the sync signals operate at much lower frequencies than the video signals. Additionally, any power levels that may be required in the pinout of a video port should be significantly filtered.

A major concern for maintaining constant trace impedance between the video generator and I/O connector lies in how the traces are routed. This routing area generally contains an absence of power planes under the RGB traces to prevent power plane noise from corrupting these low-voltage analog signals. A simple design technique to maintain impedance control is to alter the width of the trace (change impedance) if this trace has to exhibit impedance characteristics that are different from the general traces on the board. This trace width change is applicable when referenced to an image plane. Impedance control of microstrip and stripline configurations is discussed in Chapter 4. Maintain a constant impedance for video traces routed on all layers.

It is essential that the layout establish a "quiet" analog ground zone through appropriate partitioning, moating or isolation as may be required. Locate all analog traces and components exclusively in the analog section to prevent noisy digital logic and power planes from corrupting the analog components. When designing the moat for analog and digital partition, use the 20-H rule for the power plane as detailed in Fig. 5.12.

Note that certain vendors of video controllers (digital-analog converters with random access memory, or RAMDACs) tie analog ground to digital ground within the package. If the RAMDAC chosen has these two grounds connected, it is imperative that a solid ground plane be used for both analog ground and digital ground with the analog zone partitioned away (through moats with a bridge) from the digital zone. Do not use the ferrite bead-on-lead to isolate or connect grounds with this particular type of RAMDAC. Rather, connect the constant ground plane (moated to protect analog) together through a bridge. Other vendors have designed their RAMDACs for pure isolation between analog ground and digital ground through charge-coupled devices (CCDs) within the package. For these parts, use the ferrite bead-on-lead technique to isolate or connect grounds as recommended by the RAMDAC manufacturer.

Do not violate (cross) the moat with placement of any component or trace that physically resides exclusively in either the digital or analog section. Use of the 20-H rule assists in implementing this technique. It is also recommended that the surrounding planes be at ground reference rather than at voltage to minimize cross-coupling. This is shown in Fig. 5.12 for a multilayer stackup. For two-sided boards, follow the same guidelines but with extra attention to component placement and trace routing to prevent noise coupling between the analog and digital sections. The video filter discrete components must always be located adjacent to the I/O connector, with minimal possible trace length.

Certain analog monitors (monochrome) use a single coax from the I/O connector to the monitor. The braid (shield) of the coax is not an RF shield but a video signal return path. This braid (shield) may be bonded to chassis ground if long cable length ground shifts are not of concern, and it obviously must be connected directly to the video controller. If the coax braid shield is not bonded to chassis, always provide an ac shunt (high self-resonant frequency bypass capacitor) between the braid (shield) of the coax and system chassis ground to remove RF shield currents that may exist on the coax shield. Remember, attempts to isolate the coax "shield" from chassis will cause the analog ground of the video section to be "taken off" the board and be extended along the length of the video cable. In the absence of full optical-type or transformer-type isolation, the extension of "floating" analog will leave the analog devices highly vulnerable to static discharge and other coupled currents on the cable.

For PCBs with a digital video interface, use signal line filtering that is compatible with functionality on high-threat signals. Include filtering for all other traces as they relate to the rest of the board in addition to maintaining constant impedance control. Place the output section from the video generator to the I/O connector in an isolated (quiet) area. RF ground (bond) the I/O connector and cable shield directly to chassis ground immediately at the exit point, especially if the cable braid (shield) is not used as signal return.

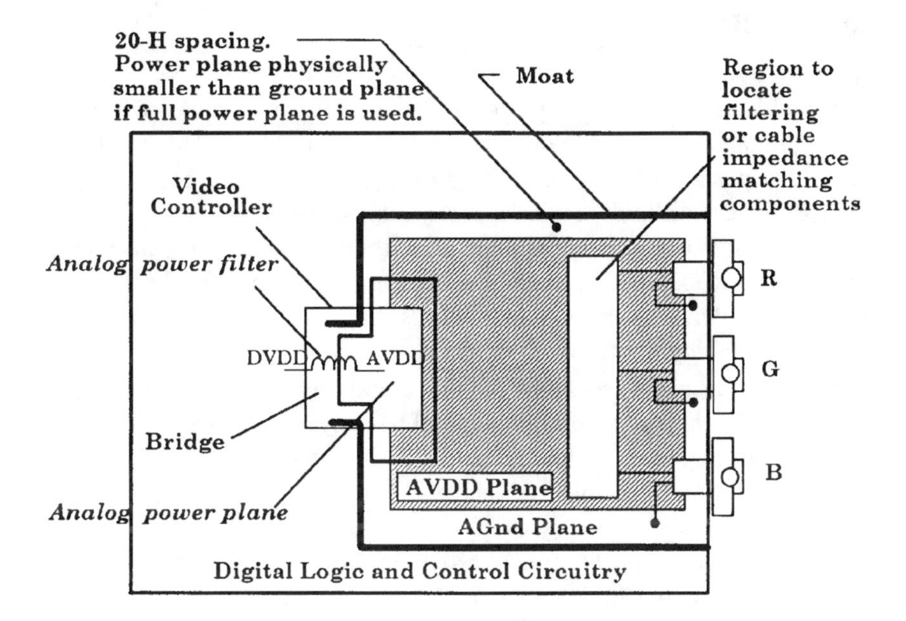

Layer Stackup (3-Dimensional Partition)

Digital traces		Layer 1
Digital ground		Layer 2
Digital traces	Dead zone	Layer 3
Digital power	Analog power	Layer 4
Digital ground	Analog ground	Layer 5
Digital traces	Analog traces	Layer 6

Moat

Layer Stackup

On layer 3 (Dead zone), nothing exist (no copper or traces)

NOTE: If the AGND and DGND are common internal to the video controller, then use only "one" ground plane. Do not isolate analog ground from digital ground! Certain vendors of video controllers tie analog ground to digital ground inside the package. Other vendors have separate analog "and" digital grounds with true isolation.

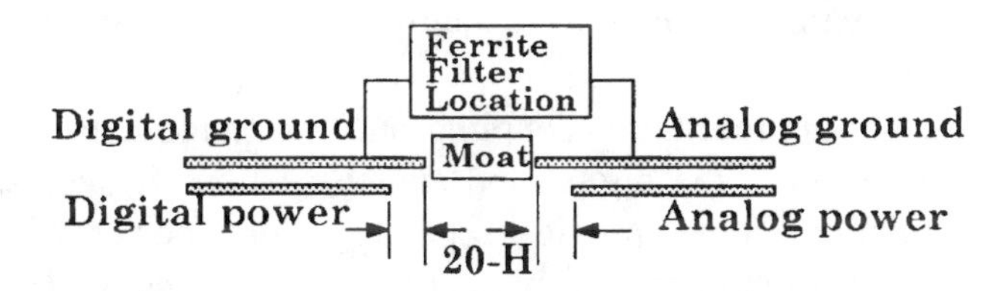

Fig. 5.12 Video circuitry layout concept

5.6 AUDIO

PCBs with audio circuitry generally contains three separate partitioned areas: digital section, analog control, and audio interface. This multilevel functional partitioning is applicable to four or more layer board stackup and can be very difficult in two-layer assemblies. Most two-layer boards do not contain moats because it is impossible to implement split planes when planes do not exist. It is possible, however, to form moat-line ground boundary separations in two-layer boards that envelop the functional circuit blocks, taking the form of "partitions."

Partition the analog section from the digital area using a bridge located *directly under* the audio controller for all traces. An example of this multilevel partitioning is shown in Fig. 5.13. No other traces can cross the moat directly or indirectly. "Isolation" between functional partitions, with fully moated boundaries, may only exist through the incorporation of isolation coupling devices such as optical isolators and transformers. In unique conditions, even power isolation may be required through transformer coupled dc-dc converters.

When designing the moat structure for both analog and digital power and ground, route power and ground traces between the digital and analog section in the immediate area adjacent to the audio controller. Apply appropriate filtering to power at these locations. If a common ground plane is used for both analog and digital components, only the power plane is to be moated (isolated). Use the 20-H rule as shown in Fig. 5.13. Analog power is provided through a ferrite bead-on-lead or inductor routed on Layer 4 as either a trace or plane.

All interconnect traces must traverse through this bridge (or isolation device) and directly *adjacent to* a solid ground (image) plane. Violation of any trace over the moat (not traveling through the bridge or isolation device) separating the analog-to-digital partition can cause digital switching noise to be injected from the digital section into the analog section. Even "white" noise may be exacerbated in this manner. White noise is random noise that has a constant energy per unit bandwidth at every frequency in the range of interest. It includes power supply and system noise and is usually heard as 50/60-cycle hum with digital transient noise spikes.

Use bandwidth-appropriate decoupling and bypassing capacitors for all components in both the digital and analog section. If necessary to minimize regulation ripple, provide separate analog power with adequately bypassed three-terminal regulator or dc-dc converters (not shown in Fig. 5.13). Such regulators must be sufficiently decoupled to

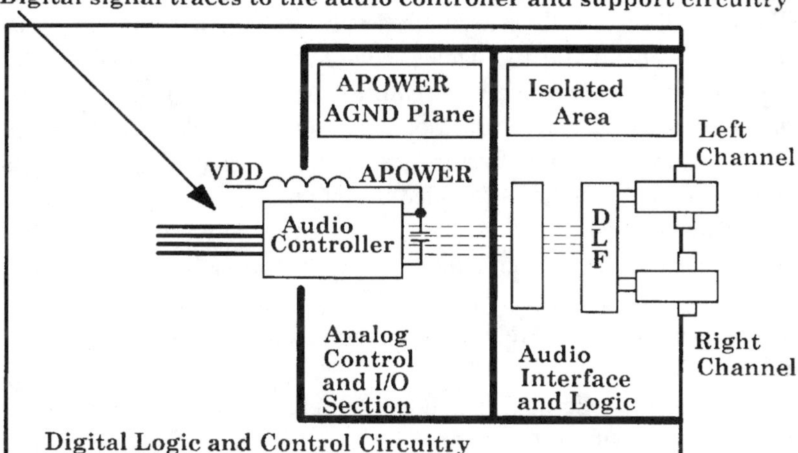

Note: DLF refers to data line filters (common-mode chokes)

Layer Stackup (3-Dimensional Partition)

Digital traces to audio controller		Analog traces	Layer 1
Digital ground		Analog ground	Layer 2
Digital traces	Dead zone	Analog traces	Layer 3
Digital power	Analog power		Layer 4
Digital ground	Analog ground	(Digital ground optional if required by controller)	Layer 5
Digital traces	Analog traces	(Digital traces if ground plane is digital)	Layer 6

Moat (D-A partition) — Moat

Note: Moats exists where shown on split planes.
On layer 3, nothing exist in the dead zone (no copper or traces)
Dead zone is required to keep analog traces from being adjacent to a digital plane. All signal traces must be adjacent to a power or ground plane.

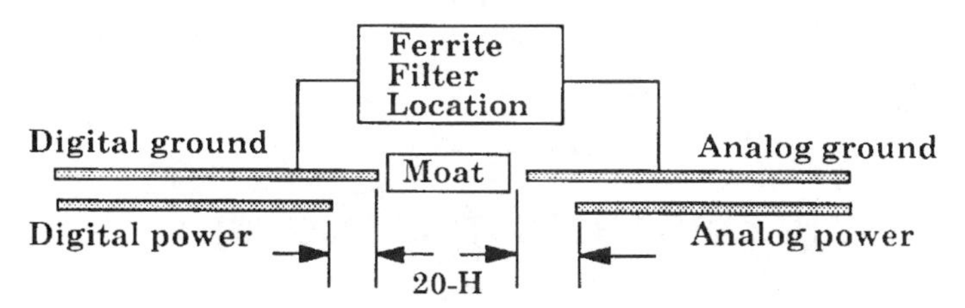

Fig. 5.13 Audio circuitry layout concept

prevent injection of power supply or digital switching noise into the analog section. Route this analog power trace through the single bridge (layer 4) to the power plane in the analog section, with the ground bridge on layer 5.

The audio interface is treated differently from both the digital logic and analog control section. To prevent chassis RF digital switching noise from coupling onto the audio I/O cables, *complete isolation of digital system power and ground planes may be required.* An audio cable may consist of a wire pair for signal and return. Coaxial braid, if used internal to an audio cable, is actually the signal return rather than RF shield ground. If this type of two-wire cabling is provided, isolate the audio I/O interface connector from the rest of the PCB with a moat, isolation devices, and the main chassis by an insulated audio connector. Isolation methods must be adequate to protect against external ESD events. If not, additional ESD protection measures that retain the desired isolation impedances will be required.

Use common-mode data line filters to remove digital common-mode RF currents present on the audio traces in addition to providing an additional set of data line filters between the analog section and audio interface. Do not provide a ground choke or inductor to reference "analog" ground to "audio" ground. A ground choke or inductor places inductance into the ground circuit. This inductance may cause ground-noise voltage to be passed from a noisy part of the board into a quiet (or clean) audio section. Place the filter circuit, either inductor or inductor and capacitor (in series), directly at the point of entry.

Locate all analog traces and components in the isolated analog planes to prevent coupling of the digital planes to the analog section. This is illustrated with a "dead zone" on layer 3 in Fig. 5.13. Do not violate the moat with placement of any component or routing of any analog traces. Use of the 20-H rule may also be required in the audio section of the PCB. It is preferred that the surrounding planes be at ground reference rather than power to minimize cross-coupling.

If the audio external interface is composed of a two-wire differential pair (for signal high and low) and a "guard" overshield, ground the overshield to chassis directly.

5.7 ENERGY HAZARD PROTECTION (FUSING)

Certain PCBs provide ac or dc power to external interconnects such as keyboards, external SCSI terminations, Ethernet MAU (AUI), optical bypass switches for FDDI applications, remote sensors, and the like. This external voltage requirement falls within the realm of product safety. Safety agencies consider the availability of an external voltage as being potentially hazardous, and certain designs features are mandatory for product approval. The requirements for externally provided power from the main PCB must be considered under the following conditions, as required in Section 5.4.9 of EN 60 950, Product Safety Standard for Information Technology Equipment (Section identical to UL 1950 and CSA C22.2-#950).

1. A circuit operating at no more than 42.2 V peak shall be inherently limited such that the output current under any condition of load, including short circuits, is not more than 8 A after 1 minute of operation.
2. Circuits operating at no more than 42.2 V peak must be provided with a protective device rated 5.0 A if the open-circuit voltage is 0–21.2 V or 3.2 A if the open-circuit voltage is greater than 21.2 V and less than or equal to 42.2 V.

Any ac or dc power voltage that leaves a PCB through an external connector must be current limited or have an in-line fuse per product safety requirements. Fusing may be implemented by a cartridge fuse, a thermal device, a series resistor, a positive temperature coefficient (PTC) device, or a pico fuse. Cartridge fuses are not easy to replace. Use of PTCs (a thermal fuse that will disable the voltage under fault conditions and re-establish connection when the fault is removed) is recommended. Because they are "self-restoring," PTCs do not have to be replaced after a fault has occurred. PTCs must be safety agency approved. This approval is mandatory as PTCs are relied upon for minimizing risk of electric shock. A failure of this device is not permitted under any condition.

The maximum current allowed through an external connector is shown below.

Open Circuit Voltage	*Fuse*	*Maximum Short Circuit Current Limit*
0.0 to 21.2 V	5.0 A	8 A (after 1 minute)
21.3 to 42.4 V	3.2 A	8 A (after 1 minute)

The reader is cautioned regulations such as these fall into the category of "product safety" and are subject to periodic revision by various domestic and international agencies, which are also responsible for enforcement. Although the above information is offered as existing examples, the reader is urged to determine the actual parameters that are required at the time of design release!

5.8 CREEPAGE AND CLEARANCE DISTANCES

A major concern of international product safety agencies is the risk of electric shock. Although creepage and clearance is not directly associated with EMC compliance, a discussion of this requirement is pertinent.

The main reason why creepage and clearance distance is of concern is that ac or high-voltage traces may exist on a PCB and be subject to an abnormal failure condition. Failures include primary-to-secondary, primary-to-ground, and primary-to-primary modes. To prevent a shock hazard due to an abnormal failure, route traces with a specific amount of spacing (distance) between high-energy (voltage) traces and secondary or ground circuits. This requirement is especially critical in power supplies and related circuitry.

When routing ac voltage traces, use sufficient trace width and spacing to comply with creepage and clearance distances described by regulatory agencies. A summary of creepage and clearance is illustrated in Fig. 5.14. Creepage and clearance distances are defined in all international product safety standards.

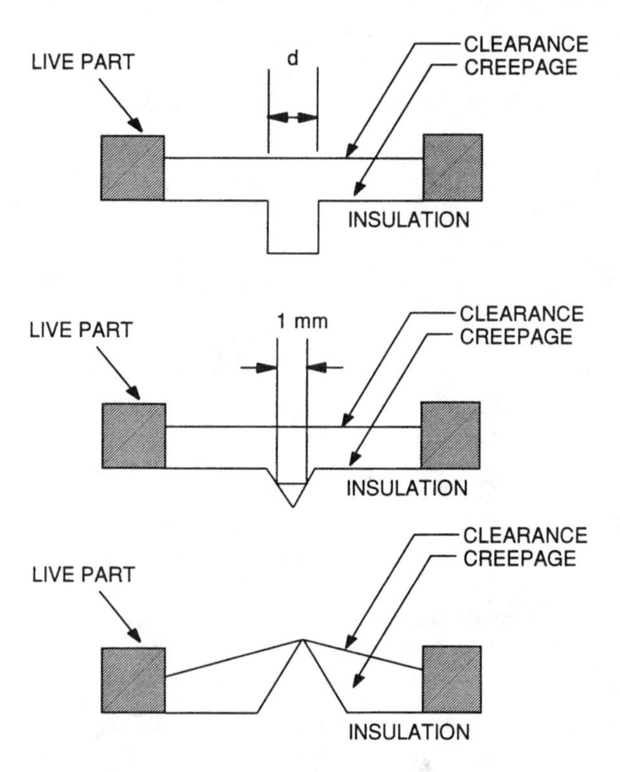

Fig. 5.14 Creepage and clearance distances

- *Creepage* is the shortest path between two conductive parts, or between a conductive part and the bounding surface of the equipment, measured along the surface of the insulation.
- *Clearance* is the shortest distance between two conductive parts, or between a conductive part and the bounding surface of the equipment, measured through air.
- *Bounding surface* is the outer surface of the electrical enclosure, considered as if metal foil were pressed into contact with the accessible surface of insulation material.

Both creepage and clearance refer to the distance spacing between traces on the PCB, between traces and/or components, and between traces and/or components and a bounding surface. Under high-voltage or abnormal fault conditions, arcing may occur across the conformal coating on the PCB, or between traces and the bounding surface. As a result, high levels of voltage and current that may be present, allow for risk of a fire hazard or risk of electric shock. International product safety standards specify minimum clearance and creepage distances for various types of equipment by working (operational) voltage range, insulation class, and degree of pollution. The reader must refer to the safety standards applicable to the product category for specific creepage and clearance dimensions.

5.9 REFERENCE

1. Drewniak, J.L., T.H. Hubing, and T.P. Van Doren. 1994. Investigation of fundamental mechanisms of common-mode radiation from PCBs with attached cables. *Proceedings of the IEEE International Symposium on Electromagnetic Compatibility*. New York: IEEE, 110–115.

6

Electrostatic Discharge Protection

6.1 BASICS

Printed circuit boards must incorporate protection against electrostatic discharge (ESD) events that might enter at I/O signal and electrical connection points. The goal is to prevent component or system failures due to externally sourced ESD impulses that may be propagated through both radiated and conducted mechanisms.

ESD pulses are generated by humans, furniture, and simple materials such as paper or plastic. The pulse may travel through multiple coupling paths including circuits, grounds, and even radiated as transient electromagnetic fields. ESD creates multiple failure modes including damage, upset, lockup, and latent failures.

An ESD event starts with a very slow buildup of energy (seconds or minutes) which is then stored in the capacitance of a structure (e.g., a human body, furniture, or unconnected cable). This charge is followed by a very rapid breakdown (typically within nanoseconds). With this pulse in the nanosecond range, the discharged energy can produce EMI in the frequency range of hundreds of megahertz to even beyond one gigahertz. An ESD event from a human can exhibit rise times ranging from approximately 200 ps to greater than 10 ns, with peak impulse currents from a few amperes to greater than 30 A [3]. Due to its high speed and high frequency spectral level distribution, ESD energy can damage circuits, bounce grounds, and even cause upsets through electromagnetic coupling.

A detailed discussion on waveforms, equivalent ESD circuits of humans, furniture, and other materials is beyond the scope of this text. Many excellent references are provided in the bibliography for those interested in mathematical and technical analysis and presentation.

Before providing design techniques to prevent ESD damage in PCBs, the ESD event itself must be understood. This will make the design techniques presented later in this chapter easier to understand and implement.

ESD introduced through a conductive transfer mode must be considered in terms of current flow, not voltage. It is like a burst dam—it is the water flow that does all the damage, not the pressure that was behind the dam before it burst. The voltage is merely a convenient metric of the electrostatic "pressure" before the ESD event occurs. ESD introduced through an indirect or radiated means must be viewed with respect to electric fields, derived from both the voltage and impulse current components, and additionally in terms of full-effect electromagnetic fields.

In addition to current levels, the ESD rise time is also important. ESD is a very fast transient. Two parameters are of great concern: peak level and rate of change (dI/dt). In the EMI world, rise times are equated to an equivalent EMI frequency based on the Fourier transform, which relates time domain signals to frequency domain components as shown by Eq. (6.1).

$$f = \frac{1}{\pi t_r} \tag{6.1}$$

where t_r = rise time. Given this discussion, a typical 1 ns rise time exhibits a spectrum of over 300 MHz. As a result, EMI immunity (not dc) design techniques are required.

There are four basic failure modes of ESD related to PCBs. These are described as [1]

1. *Upset or damage caused by ESD current flowing directly through a vulnerable circuit.* This relates to any current injected into the pin of a component that causes permanent failure. Through this mode, direct connections from a component to the outside environment (i.e., keyboard) can carry damaging ESD pulses. Even a small amount of series resistance or shunt capacitance in these circuits can limit the ESD current through the IC, although the acceptance value is specific to each IC type.
2. *Upset or damage caused by ESD current flowing in the ground circuit.* Most circuit designers assume that the circuit ground has a low impedance. With 1 ns rise times, the ground impedance may not be low. Hence the ground will "bounce." The usual result is an upset. Ground bounce can also drive CMOS circuits into *latch-up*. Latch-up is a situation where the ESD doesn't actually do the damage—it just sets things up so that the power supply can destroy the

part or, at best, that the circuit becomes non-functional without a power cycle reset.

3. *Upset caused by electromagnetic field coupling*. This effect usually does not cause damage (although there have been reports of damage to very high-impedance component devices), because typically only a small fraction of the ESD energy is coupled into the vulnerable circuit. This failure mode depends heavily on the rise time (dI/dt), circuit loop areas, and presence or absence of shielding. This effect is often called the *indirect coupling mode*. Electromagnetic field sources do not have to be very close to cause problems to a sensitive circuit.
4. *Upset caused by the pre-discharged (static) electric field*. This failure mode is not as common as the others. This mode appears in very sensitive, high-impedance circuits.

A pre-discharged (static) electric field is caused by stripping electrons from one object (resulting in a positive charge) and depositing these electrons on another object (resulting in a negative charge). In a conductor, charges recombine almost instantly, while in an insulator, the charges can remain separate. In an insulator, it may be a long time before significant charge recombination occurs and, consequently, a voltage builds up. If the voltage becomes large enough, a rapid breakdown occurs through the air or insulator, creating the familiar ESD arc or spark [1].

Because ESD is transient in nature, fast digital circuits are more prone to ESD upsets than are slow analog or, for that matter, low-bandwidth digital device circuits. In fact, ESD rarely upsets the functionality of analog circuits. However, both analog and digital circuits are vulnerable to ESD damage from a direct discharge. Digital circuits with edge rates faster than 3 ns are particularly vulnerable because they can be fooled by phantom ESD pulses. As a result, digital circuits are more vulnerable than older circuits with slower edge rates.

Several commonly used design techniques for ESD protection that may be implemented on a PCB for high-level pulse suppression include the following:

1. *Spark gaps*. These are sharply pointed triangles aimed at each other, with the pointed tips separated by a maximum of 10 mils and minimum of 6 mils (one mil = 1/1000 inch). One triangle is part of the ground plane, and the other is situated on a signal trace. The triangles are not components but are made up of traces of copper on the PCB layout. These triangle spark gaps must be placed only on

the component (top) layer of the PCB, with no solder mask. The only signals or connections exempt from use of spark gaps are those determined by safety agencies to require dielectric (hi-pot) testing. Spark gaps are shown in Fig. 6.1. (Hermetically sealed spark gaps are also available as component devices from manufacturers and are generally too slow in responding to an ESD event.)

2. *High-voltage capacitors*. These disc-ceramic capacitors must be rated at 1500 V (1 KV) minimum. Lower-voltage capacitors may be damaged by the first occurrence of an ESD pulse. This capacitor must be located immediately adjacent to the I/O connector. (Note that a capacitor used as an ESD clamping device will be subjected to a ratio amplitude, viewed as the derivative of the clamping capacitor impedance versus the ESD capacitive source value and impedance. Given this proportional effect, lower-amplitude capacitors should be used if they can withstand the ESD impulse stress current.)
3. *Tranzorbs*™.* These are semiconductor devices specifically designed for transient voltage suppression applications. They have the advantage of a stable and fast time constant to avalanche, and a stable clamping level after avalanche.

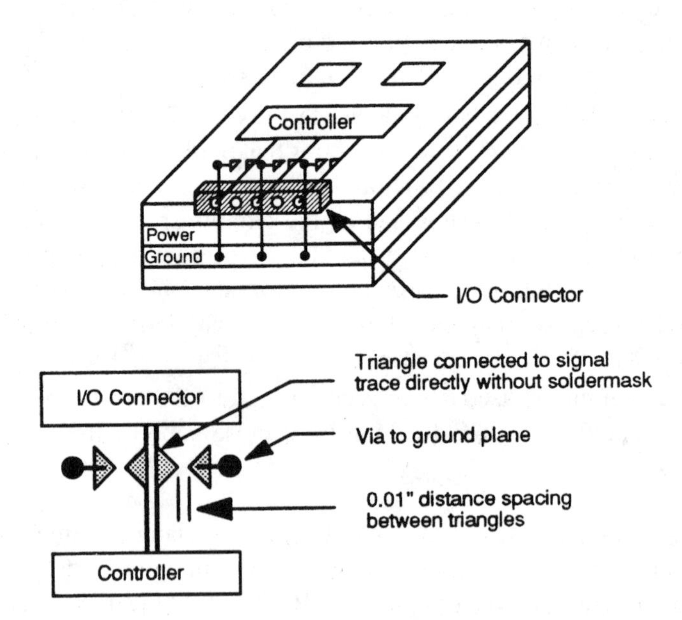

Fig. 6.1 Spark gap implementation

*Tranzorb is a trademark of General Semiconductor.

4. *LC filters.* An LC filter is a combination of an inductor (which may include the inductor inside a data line filter) and a capacitor to ground. This constitutes a lowpass LC filter that prevents high-frequency ESD energy from entering the system. The inductor presents a high-impedance source to the pulse, thus attenuating the impulse energy that enters the system. The capacitor, located on the input side of the inductor (not the output or I/O side) will shunt high-frequency ESD spectral level components to ground. An additional benefit of this circuit combination is enhancement of radiated EMI noise suppression (at the expense of rolling off the edges of the data signals and possibly compromising signal integrity). This LC combination may consist of a multi-line common-mode data line filter in a single package with a discrete capacitor.

Air-type spark gaps, if used in an area that is subject to frequent ESD events, will eventually break down and create carbon tracking when an arc occurs between the two triangle points. Carbon is a conductive material that could eventually short out the spark gap. The net result is that the signal trace can be permanently shorted (or have an undesired resistance) to ground. As a result of carbon tracking, use of air-gap type spark gaps is not recommended in open-air installations where ESD events frequently might be anticipated. However, hermetically sealed spark-gap component devices are usually stable over time.

To illustrate the intensity of ESD, rise times of 500 ps or faster have been reported by researchers. Consider an ESD rise time of 500 ps with tens of amperes of peak current. This translates to equivalent slew rates or giga-amperes per second across the circuit interface!

Two primary ESD design implementation techniques for PCB layout are commonly used. Several areas of concerns exist for each technique, as described below [2]. Generally, these techniques are the same as those described for good EMI design.

Minimizing loop areas

Identify areas where loop currents can exist. This includes distance spacing between components, I/O connectors, and component/power planes. The following techniques are useful for minimizing loop areas.

- Keep all power and ground traces (single- or double-sided PCBs) close together if power planes are not used. This includes adding additional feedthroughs to connect both power and ground traces together in more locations, thus minimizing loop currents.

- Keep signal lines as close as possible to ground lines, ground planes, and circuits. This is illustrated in Fig. 6.2 for both single- and double-sided boards.
- Use bypass capacitors with a high self-resonant frequency between power and ground. These bypass capacitors must have as low an equivalent series inductance (ESL) and equivalent series resistance (ESR) as possible. Frequent use of bypass capacitors reduces loop areas of the power and ground planes for low-frequency, high-level pulses. For higher-frequency ESD events, standard capacitors may become less effective due to the capacitors' internal stray inductance and the interconnect trace inductance to the component or ground stitch connection.
- Keep trace lengths short. Traces act as antennas (for both radiated and susceptibility concerns) at various wavelength multiples of a particular frequency. Thus, a long trace line will be susceptible to more ESD energy. Group components as closely as possible in areas susceptible to ESD, and minimize trace length!

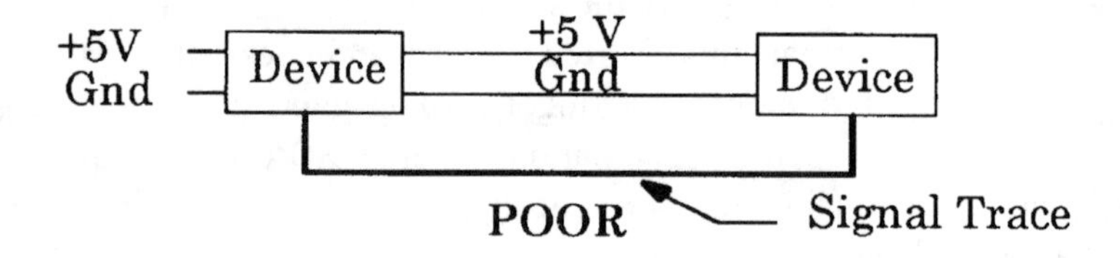

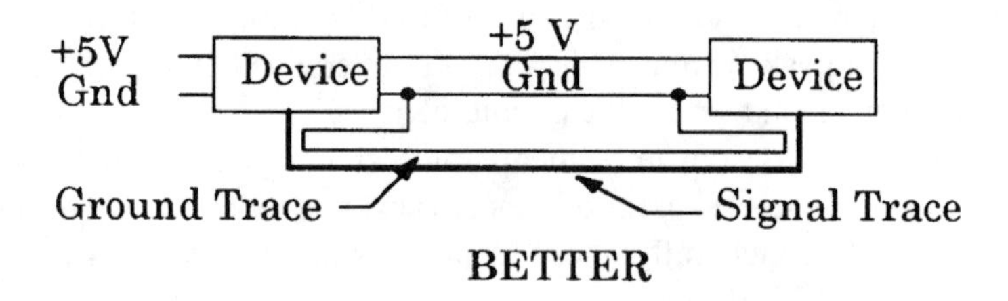

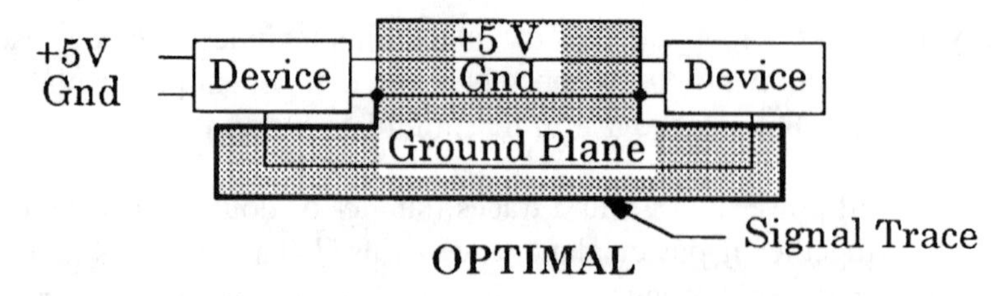

Fig. 6.2 Routing signal traces close to ground

- Fill in both top and bottom layers of the PCB with as much ground plane (ground fill) as possible in areas without circuits or components. This localized ground plane reduces ESD effects by acting as a low-impedance path to chassis ground or system ground. This low-impedance path conducts the high-energy pulse to ground rather than into signal lines or components. The one disadvantage of using fill areas as a ground is that an ESD pulse could be discharged into the ground system of the product, thus causing potential damage to components, nonfunctionality, or system glitches. Ground fills must be connected to ground planes at frequent intervals.
- Coupling between ground and power subsystems must be tightly controlled. This is accomplished by keeping both power and ground traces close together (or power planes adjacent to each other). Tight coupling is also achieved by using high self-resonant frequency bypass capacitors between power traces (or planes). Current injection problems from a high-level pulse are also reduced. Parallel bypassing is effective when capacitors are chosen for both high- and low-frequency ESD events.
- Implement moating and isolation between ESD-sensitive components from other functional areas. Moating and isolation are discussed in Chapter 5. Isolation prevents an ESD event from being transmitted or coupled from one functional section to another.
- Make all chassis ground connections low impedance. Locate these ground connections at positions where they will conduct ESD impulse energy away from sensitive circuits rather than through them! A low-impedance path to chassis ground will source the ESD event away from susceptible circuits without arcing between the traces. Provide a chassis ground trace with a length-to-width ratio of 4:1 or less (compared to the width of the signal trace). Keep this ground trace as short as possible.
- Ground planes internal to the PCB should surround every plated through-hole to minimize the creation of loop areas between circuits and traces.
- Provide transient protection such as zener diodes or Tranzorbs™. These devices must be fast enough to react to the ESD event and designed to dissipate a lot of energy in a short time. For optimal performance, keep the leads short. A few millimeters of lead length will degrade the performance of this protective device. High-frequency capacitors may be used in parallel with these

devices to slow down the ESD pulse, thereby giving the protection device time to clamp.

- Ground transient protection devices to chassis ground, not circuit ground. ESD events contain large amounts of current. Placing high current levels on ground planes can cause serious ground bounce and possible component failure.
- Ferrite material (beads and filters) provide excellent attenuation of ESD currents, in addition to providing EMI protection for radiated emissions. Hence, two suppression features with the use of one device are obtained.
- Use of multilayer PCBs provides 10 to 100 times improvement over two-layer boards for protection against electromagnetic fields of indirect ESD. Locate the first ground plane as close to the signal routing plane as possible so ESD cancellation can occur if an ESD event reaches the traces.

Guard band implementation

Guard bands are different from the guard traces detailed in Chapter 4. They are intended to minimize ESD risk to the boards when they are being handled by the edges. To prevent radiated or conductive coupling into components from an ESD event not related to I/O interconnects, place a 3.2 mm (1/8") guard band around all four edges of the PCB, both component (top layer) and circuit (bottom layer) side. Connect this guard band to all ground planes by vias every 1/2" around the entire PCB. This low-impedance ground connection from the guard band to the ground plane will sink ESD energy through a low-impedance path to chassis ground or earth ground. If a moat is used on the PCB traversing to the edge of the board, break the guard band at the partition break. This break in the guard band will not degrade EMI or ESD performance; however, if a moat does not break this guard band, the potential for RF catastrophic failure exists. Violating a moat with a guard band, even at the end of a moat, allows for parasitic capacitance to be generated between the planes separated by the moated area, thus causing susceptibility problems for both EMI and ESD. Details of how to implement a guard band are shown in Fig. 6.3.

When using guard bands, all signal traces on the PCB should be kept away from the edge of the PCB by a distance equal to the trace height above the ground plane. For example, if the distance spacing between the trace layer and the ground plane is 0.006 inches, then the signal trace must be routed 0.006 inches inward from the guard band.

Depending on the layout, the priority for component placement to protect circuits from ESD consists of one of the following sequences:

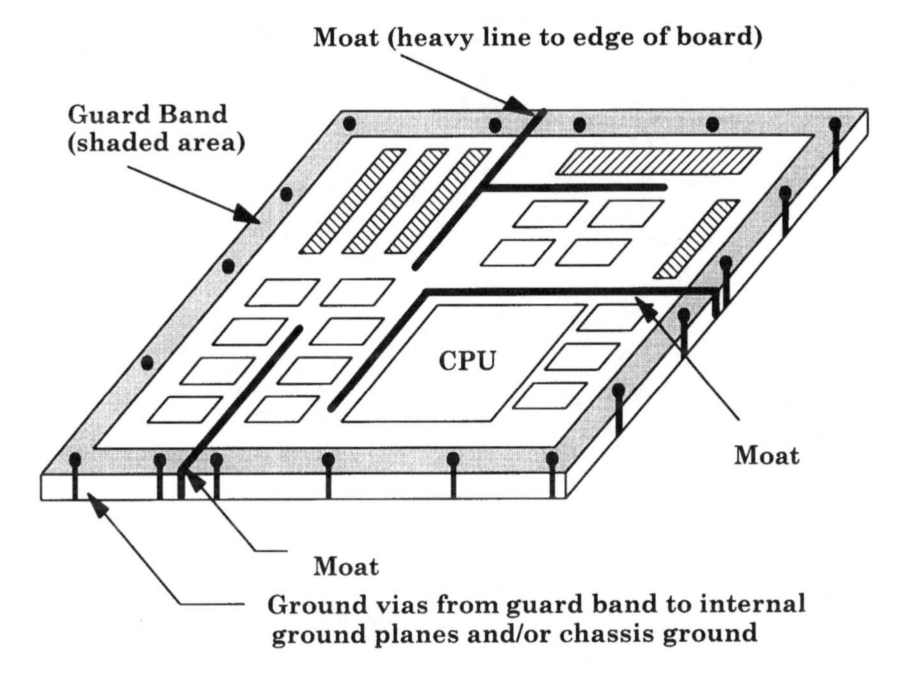

Fig. 6.3 ESD guard band implementation for handling

1. I/O connector, high-voltage ESD filter capacitor, the ESD diodes (Tranzorbs™), a bypass capacitor,* I/O controller
2. I/O connector, data line filter (with a moat traveling through the middle), bypass capacitor, I/O controller
3. I/O connector, spark gap, data line filter* (with a moat traveling through the middle), bypass capacitor, I/O controller

6.2 REFERENCES

1. Gerke, D., and W. Kimmel. 1994. The designers guide to electromagnetic compatibility. *EDN* (January 20).
2. Boxleitner, W. 1988. *Electrostatic Discharge and Electronic Equipment.* Piscataway, N.J.: IEEE Press.
3. American National Standard, Guide for Electromagnetic Discharge Test Methodologies and Criteria for Electronic Equipment, ANSI C63.16–1993.

* optional components

7

Backplanes and Daughter Cards

This chapter provides an overview on PCB layout techniques for both backplanes and daughter cards. Several concerns exist when designing backplanes and daughter cards. These include trace impedance, assembly construction, trace termination, signal routing, crosstalk, and trace length. All design rules and techniques applicable to PCB layout presented in earlier chapters also apply to backplanes, daughter cards, and motherboards with adapter slots.

7.1 BASICS

Before designing a backplane, first determine the connector pin assignment. Assigning the connector pinout at this stage will (1) assist in prevention of crosstalk, (2) reduce radiated emissions, (3) provide signal quality enhancement, and (4) help maintain proper ground loop control. A backplane is essentially the *freeway* of signal flow between interface circuits and daughter cards. As such, special consideration must be taken to ensure proper impedance control and termination of signal lines exist.

Impedance matching between the backplane and daughter cards is a primary requirement in high-technology products. Generally, signal traces are longer in a backplane than equivalent traces on daughter cards. This is because a signal is usually generated on an adapter card and sent through two connectors (plug and receptacle, both of which may have different impedances), after which it travels down a trace in the backplane and then through another pair of mated connectors to a destination board. The destination board now has to receive this transmitted signal and transfer it to its appropriate receiver (load). A long transmission line may exist for this trace. Proper impedance matching maximizes signal quality while minimizing RF energy, crosstalk, and voltage (IR) drop. Impedance matching includes

- designing the appropriate trace impedances throughout the product
- selecting the backplane-to-daughter card connectors so that they will be capable of maintaining that match
- establishing the appropriate pinout for the connector to facilitate the match while providing sufficient grounds to minimize ground shifts across the connectors
- selecting connector intervals along the backplane that are appropriate to minimize "internal reflections" commensurate with the signal edge times
- locating the receivers/drivers/transceivers on the daughter cards to minimize "stub reflections," considering the edge times

Because backplanes with through-hole connector pins have large via holes where the trace enters at the connector location, a decrease in impedance is observed. Typically, the impedance, Z_0, is 50 to 70 Ω. Large vias in the backplane will drop the impedance to 30 or 40 Ω. With a lower impedance, more RF current is generated in the trace, which in turn allows more RF energy to couple to other circuits, subsystems, or free space. A reflection discontinuity is developed that affects propagation skew and signal quality.

Impedance-controlled connectors, although more expensive than traditional connectors, reduce or eliminate much of the impedance mismatch that may occur between a motherboard and adapter cards or the backplane. These connectors are available in both microstrip and stripline versions. Both versions provide a very high pin count and exhibit reduced ground bounce. The net result is higher performance of the interface and enhanced signal quality for reliable operation. With proper design techniques as presented in earlier chapters, higher-quality data and clock signal traces are possible in backplanes and daughter cards.

7.2 TRACES AND PARTITIONS

Differential-mode RF currents in a system can be radiated at the discontinuities presented by the physical connection between the backplane and motherboard or between a motherboard and daughter card. This interconnect is performed by cables or connectors. Differential-mode RF current sources are many, and the unpleasant fact is that EMC compliance is often neglected in the design of backplanes, daughter cards, and interconnects.

Backplanes usually consist of multiple clock and signal traces sharing a single common ground-return plane. When ground pins are assigned throughout the entire length of the connector, minimize loop areas to prevent high levels of RF currents from being coupled to other components and subsystems. Loop control on PCBs can be maintained using a multilayer stackup with both a ground return plane and alternating ground pins between clock and signal lines on the interface connector. An example of this is shown in Fig. 7.1. In prioritizing pin designations and signal locations in backplane-daughter card connectors, first allocate the highest-frequency, fastest-edge signals to the pin rows with the lengths of the shortest pins, and the slowest-edge, lowest-frequency signals to pins having the longest lengths of the mated connector pairs.

Some applications require use of an ac *chassis* ground plane in addition to the *logic* ground planes. This ac *chassis* ground plane has no direct connection to the other logic ground planes except through the interplane capacitance provided distributively between chassis plane and signal ground planes (which may be augmented by an array of 1 nF capacitors around the perimeter) within the structure of the backplane. The ac chassis plane is then directly connected to the main chassis ground at frequent intervals around the perimeter. The combination of the distributive and

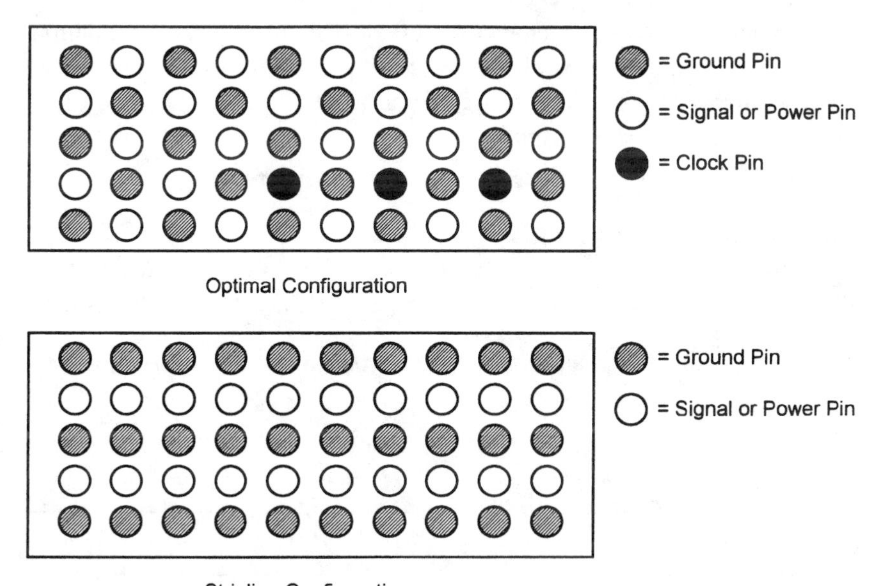

Fig. 7.1 Recommended layout of ground return pins for interconnects

component bypass capacitance between the chassis plane and the ground plane shunts RF currents generated in the internal logic ground plane to chassis ground without setting up direct "dc" connections between power currents and chassis. Use of ac chassis planes is primarily effective when used as a "Faraday shield partition" internal to the backplane. Such partitions allow for a separation of EMI fields from, for example, highest-frequency system bus activity (routed in layers closest to the system cards) and lower-frequency bus activity (routed in layers closest to "back" cards). The ac chassis plane is actually used as a "midplane" in a common card cage.

When designing backplanes or interconnects, special consideration must be given to all clock and periodic signal traces. A clock trace *must always have an adjacent ground return pin on all sides of the connector or interface, no exception*. This is shown in the lower portion (optimal configuration) of Fig. 7.1.

7.2.1 60 and 100 Ω trace impedance examples

The following examples of trace widths are given for both G-10 and FR-4 PCB material for two- and four-layer boards. These examples represent typical calculations that must be made when designing backplanes. Table 7.1 illustrates this, based on Eqs. (4.2) through (4.8), from Chapter 4. For larger stackup boards, the same concepts apply.

To achieve a trace impedance of 60 or 100 Ω, use the calculations below. Traces routed in parallel must be a minimum of two trace widths away as measured from the inside of one trace to the inside edge of the parallel trace (as derived from the geometry of the 3-W rule) for stripline or microstrip.

Four-layer board example

For a 60 Ω impedance PCB or backplane, the ratio of trace width, w, to lamination thickness, t, is typically 1.35 (w/t = 1.35). Assume the lamina-

Table 7.1 Ratio of Trace Width to Nearest Plane (Ground or Voltage), Two-Layer PCB

	Lamination Thickness (Inches)		*Trace Width (Inches)*		*Trace Separation (Inches)*	
PCB Material	*60 Ω*	*100 Ω*	*60 Ω*	*100 Ω*	*60 Ω*	*100 Ω*
G-10	0.135	0.40	0.027	0.008	0.054	0.016
FR-4	0.120	0.36	0.024	0.005	0.048	0.015

tion thickness is 0.020 inches (standard lamination thickness of a four-layer board, total stackup thickness of 0.062 inches). The trace width would be 0.027 inches ($w = 1.35 \times 0.020$). To achieve an impedance of 60 Ω, adjacent trace layers would need to be a minimum of 0.054 inches away ($0.027 \times 2 = 0.054$).

Impedance selection

Trace impedance optimization for backplanes depends on the edge rate of the signals (not frequency) compared to the propagational time interval between connectors, or the allowed separation permitted between unpopulated connector positions. If these intervals approach approximately 10 percent of the edge rate of the signals (given an FR-4 propagational time constant of approximately 1.7 ns/ft), the fully populated loaded bus impedance value should be used. Frequently, for a 15 to 19 slot backplane, values found are between 30 and 50 Ω.

7.3 BACKPLANE CONSTRUCTION

7.3.1 Basics

There are several basic items to remember for daughter cards and plug-in modules. These items are in addition to the information presented in Chapter 2. Consideration must be given to use of the outer layers as signal routing (microstrip) or a solid plane (stripline for trace routing). Generally, a plug-in module is positioned at a 90° angle from the main board (e.g., adapter boards for personal computers) or cardcage assembly where the module is plugged into a backplane.

Five main areas must be considered in addition to those presented in Chapter 2. These are:

1. purity of the power planes from ground bounce and high-frequency RF currents injected into the planes
2. signal quality of the bus, which may contain many parallel traces
3. impedance control and capacitive loading
4. interboard coupling of RF currents
5. field transfer coupling of daughter cards to card cage

Power plane purity

Switching noise from the power supply, radiated or conductive coupling of RF currents from other parts of the system, voltage drop (IR), and ground bounce all affect the purity of the voltage that is provided for components and daughter cards.

Contamination of the voltage plane is possible. This may affect the performance of sensitive components—especially analog and PLL circuits. Contamination may be caused by switching noise from the power supply, external RF fields, ESD, electrical overstress (EOS) events, I/O cables, interconnects with a ground reference, and peripherals along with motors and magnetic components injecting inductive switching noise into the power system (i.e., disk drives). Contamination of low-voltage-sensitive circuits may cause functional degradation.

Voltage drop in the backplane occurs when multiple boards are inserted, with one board consuming much more power at the input connector than another board inserted at another position on the backplane. An IR drop may occur between these boards that will also affect functionality and performance.

Ground bounce (sometimes called *ground shift*) is one of the main contributors in corrupting signal functionality. Ground bounce (and power bounce) is observed when many large power-consuming circuits switch all their component pins simultaneously under maximum capacitive load. Decoupling capacitors remove high-frequency RF currents injected into the power planes from components. Bulk capacitors prevent power dropout to maintain proper voltage reference. Consideration must be made related to eliminating ground bounce in a backplane above and beyond discrete capacitors provided on each adapter module. Extensive use of both bulk and decoupling capacitors must be provided for every I/O connector on the backplane. The capacitors minimize ground bounce and maintain signal purity for other boards present in the system in addition to the capacitors on individual boards. It is imperative that voltage planes be located immediately adjacent to a ground plane to reduce the parallel power plane dynamic impedance.

Signal quality of parallel overlaying traces

A concern for signal quality in a backplane lies in the large number of traces running in parallel along the length of a backplane. Two items with the greatest concern are crosstalk (Section 7.8) and ground slots (Section 7.10), both discussed later in this chapter. Trace termination must also occur if the routed length of the trace is electrically long as detailed in Chapter 4, combined with the reflection between backplane connector intervals versus the edge times of the backplane signals.

Crosstalk is often overlooked during layout. It is generally a poor practice to route traces on adjacent stripline planes in the same axis without making a careful assessment of cross-coupling possibilities. A backplane generally has primarily x-axis trace routing. Interplane coupling from two adjacent stripline routing planes may corrupt the signal quality to each mutually adjacent routing plane. If this situation occurs, the recommendation for large bus systems is to route each and every routing plane between two solid image planes, with "ground" being the preferred potential for these planes. It is preferred that two stripline planes not be mutually adjacent in extremely high-speed, high-technology, multilayer backplanes. In mid-bandwidth systems, where cross-coupling is not a serious threat to performance, parallel adjacent signal routing layers may be acceptable in a ground-signal-signal-ground stackup where the separation between the signals is $\geq Z \times$ the separation from each signal layer to its adjacent ground plane.

Impedance control and capacitive loading

When multiple boards are inserted into a backplane, the characteristic impedance of the backplane will change due to capacitive loading presented by each board. In many situations, the actual impedance of the backplane signal traces may not be controlling the desired value. The loading provided by the daughter cards will control the impedance where the signal edge times are long compared to the interval time of the slots. Signal quality concerns and impedance control for trace termination could be seriously degraded.

When designing a backplane, perform an impedance study on the load characteristics and match the board to this impedance. The specification for VME is 100 Ω; however, a fully loaded bus in reality may be 20 to 30 Ω.

If a bus on a backplane, by analysis of daughter card loading, has the ability to contain multiple impedance values (40/50/60 Ω), select the value in the middle of the range; i.e., 50 Ω. Choosing the nominal impedance for the board will minimize impedance swing in a fully loaded or no-loaded bus. In designs where the propagational intervals between the locations of daughter card occupancy (at approximately 2 to 2.2 ns/foot) become a significant percentage of signal edge times, design the trace impedance to the full-population value.

Interboard coupling of RF currents

This design consideration is generally overlooked by most system-level designers. For multiple board configurations, each board is usually considered on an individual basis, without concern for where this board is located within the backplane assembly, or if it is adjacent to another

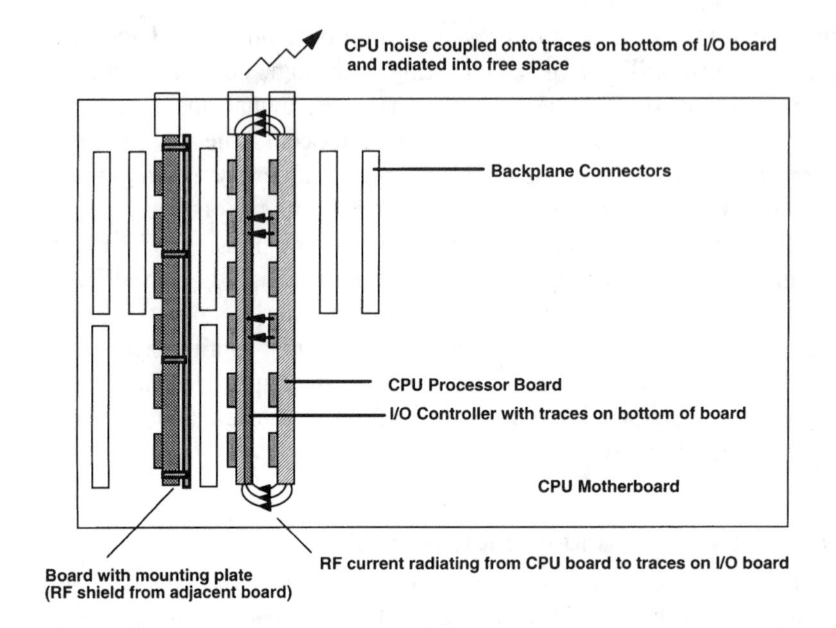

Radiated coupling between boards in a backplane. I/O controller is susceptible to RF corruption. High threat traces (clocks and I/O) cannot be routed on bottom layer of I/O board if multilayer stackup is used.

If I/O board is placed on left side of CPU, then radiated coupling could occur from components and signal traces located on top side of the adjacent CPU board.

To minimize interboard coupling, make bottom layer of I/O board a solid ground plane or a signal layer without trace routing. If not possible to have a mounting plate, a metal plate may be used.

If I/O board was located on right side of CPU, radiated coupling would not be observed and EMI compliance is assumed to exist within the product.

Fig. 7.2 Interboard radiated coupling

board with high-threat signals. An example, shown in Fig. 7.2, places a 100 MHz clock circuit on a daughter board plugged into a backplane (components and traces on the top layer). We install this board in a card-cage with an I/O board adjacent to the clock circuit board. For this situation, there usually exists a minimal trace separation distance such as 0.100 inches (0.25 cm), or 0.25 inches (0.63 cm) between boards. It is observed that RF currents from the clock board will radiate to the I/O board and cause possible noncompliance with EMI requirements or contamination of I/O logic circuitry. If the I/O board is positioned in a card slot on the opposite side of the clock board, compliance is surely met since the power and ground planes internal to the clock board will help contain RF currents on the clock board from coupling into the I/O board through radiated means.

With the situation mentioned, it is important to note that high-threat (clock or I/O) traces must be routed stripline when multilayer boards are used in a backplane. The same concern exists for daughter cards, and particularly lower-frequency external interface cards with unshielded interface cables. If adapter card designers forget interboard coupling, EMI problems may be difficult and very expensive to locate and fix. A layout technique is to design *ALL* daughter boards with the bottom layer being a ground plane without segmentation or traces of any kind. With this design technique, the probability of compliance is improved, providing that the other techniques presented in this guide are implemented for the entire assembly.

For those applications where a ground plane cannot be implemented, and internal coupling of RF current occurs from an adjacent board, use of a metal shield is required. This shield is commonly referred to as a *mounting plate* when used in a card cage assembly. This metal shield minimizes interboard coupling in a card cage environment. When taken to full implementation, this permits daughter cards of highly incompatible bandwidths or sensitivities to be intermixed randomly in any combination in a card cage without significant concern for interboard coupling. Use of this mounting plate and problems related to its use are illustrated in Fig. 7.3. An example of a daughter board with a mounting plate is also presented in Fig. 7.2.

Field transfer coupling of daughter cards to card cage

This situation is similar to interboard coupling of RF currents except that the RF fields generated from a board (components, ground loops, interconnect cables, and the like) will couple to the chassis or card cage. As a result, RF eddy currents will exist in the chassis and circulate inside the unit, creating a field distribution. This field could couple to other circuits, subsystems, interconnect cables, peripherals, and so on. One of the most significant ramifications of this field distribution is to develop a common-mode potential between the backplane and the card cage. This potential will exhibit the spectral energy signature not only of the backplane, but the daughter card as well. In addition, this field will be observed during radiated tests in the near field ($< \lambda/4$), or as an electric field at a distance greater than $\lambda/4$ at the frequency of concern. Proper implementation of suppression techniques on a PCB, and proper referencing of the backplane to the card cage to short out the distributively derived potentials, will minimize field transfer coupling between the boards to the backplane and card cage.

The "proper referencing" of the backplane to the card cage noted above takes the form of establishing a very low-impedance RF reference

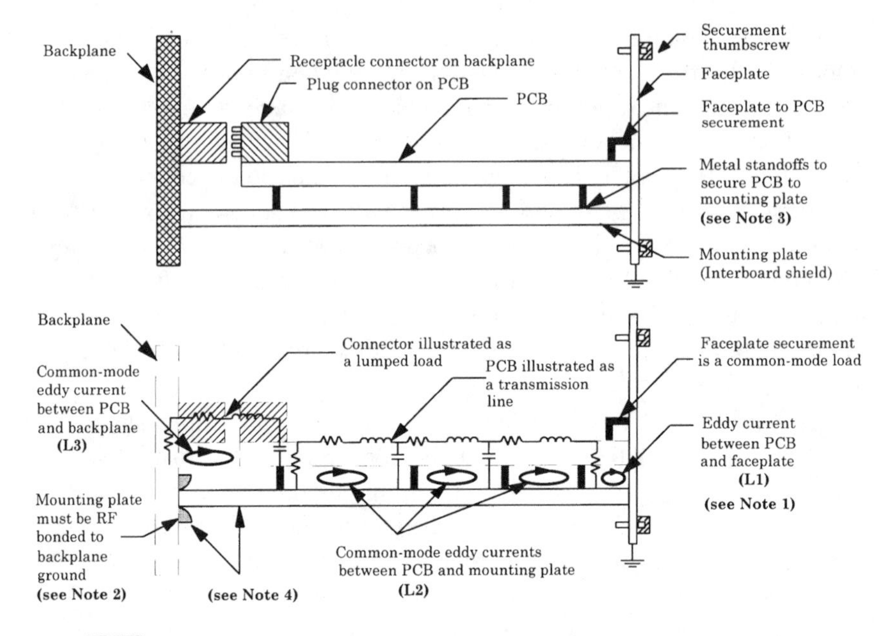

NOTES:

1. To control the common-mode loop L1, the faceplate must be RF bonded to the mounting plate and probably to the circuit board.
2. To control the common-mode loop L3, the mounting plate must be bonded to the backplane with a low-impedance connection. Without this connection, the mounting plate will couple (or transfer) common-mode eddy currents to the chassis or adjacent printed circuit boards.
3. Metal standoffs are most effective when used as ground connections (nulls) between the printed circuit board and mounting plate.
4. The common-mode "E" losses across the connector are reduced by the ground-ground short across these positions. This method significantly reduces the common-mode EMI driving the daughter card.

Fig. 7.3 Backplane interconnect impedance considerations

between the backplane and the card cage. This reference method is mandatory to short out the eddy currents developed at and by the daughter cards. These currents are coupled to the card cage through distributive transfer impedances (often in the low tens of ohms) and then attempt to "close the loop" by coupling to the backplane. If the common-mode reference impedance between the backplane and the card cage is not significantly lower than the distributive "driving source" (of the eddy currents), an RF voltage will be developed between the backplane and the card cage. This voltage will have the spectral energy profile signature not only of the backplane but of the daughter cards as well. This voltage will cause any conductors that are connected to the backplane to radiate the spectral profile—even dc wires! The spectral voltage developed in this mechanism can even contribute to interboard coupling using the backplane-to-card cage relationship as an intermediary!*

Simply, the common-mode spectral potential between the backplane and the card cage must be shorted out. This may take the form of frequently connecting the backplane ground plane to the card cage (chassis) at regular intervals around the perimeter of the backplane. Alternatively, an "ac chassis plane" could be configured in the backplane, positioned immediately adjacent to a logic ground plane, so that a distributive transfer impedance will be established between the ac chassis plane and the ground plane. This chassis plane may also serve as a "Faraday partition" within the backplane (see Section 7.2). The location of an "ac chassis plane" within the backplane must be such that it is *never* used as an image return reference for signal traces—meaning that it must be "capped" by logic ground planes. Generally, to be reasonably effective, the RF transfer impedance between the logic ground planes and the ac chassis plane must be equal to or less than 1 Ω, thereby shorting out the common-mode potential between the daughter cards-card cage-backplane-to-card cage.*

7.3.2 Number of layers

Construct the backplane with a minimum of four layers—two routing and two power planes (voltage and ground). Designing a backplane with less than four layers is not recommended. In a four-layer (or more) stackup, the outside layers are generally used for signal routes, and the inside layers for ground and voltage, respectively. The spacing between any two adjacent layers will be different if a controlled impedance stackup is required.

The reader is cautioned that the best EMI and system performance will be gained when the signal impedances are well controlled and referenced to ground planes rather than voltage planes, and that the intrinsic parallel-plane power impedance distribution is established at as low a value as is reasonably possible. To conform to these goals for a backplane having two or more signal routing layers, multiple slot positions exceeding approximately five daughter cards, and signal edge rates of faster than approximately 5 ns, the use of a simple four-layer card becomes essentially incompatible. More layers are required!

When designing six-layer or greater backplanes, follow the guidelines presented in Chapter 2. Specifically, the concept illustrated and annotated in Fig. 2.10 is preferred because the power flux is maintained away from

* These propagational mechanisms and solutions were derived and modeled by W. Michael King.

the signal flux paths. Connect the ground planes of the backplane to chassis ground at close intervals if direct connection is being used to reference the backplane for common-mode fields. Use the "ac chassis plane" referencing method if the backplane is to be referenced without direct dc connections.

7.3.3 Number of connector slots

Determine the edge rate of the fastest clock and periodic signal trace that must travel through the connections of the backplane. Calculate the maximum electrical length of these traces using Eqs. (4.14) through (4.17). If there are many connectors, measure the total physical dimension between the two farthest connectors. Remember that the backplane in real-world use may not have daughter cards populated in all slots, leaving "population gaps" in the signal flow and thereby dramatically influencing both the timing skew and the reflection diagram. Also include the distance the signal trace must travel on the daughter cards that plug into the motherboard. Perform worst-case timing analysis to determine if waveform degradation occurs when many connectors are used, or if the physical spacing is electrically long between the two end connectors and the source/load point on the daughter cards. Termination of signal lines may be required.

With many connector slots, a larger value of lumped distributed capacitance is presented to the backplane when a PCB is actually installed in the slot. With additional capacitance, degradation of signal quality occurs, sometimes to the point of nonfunctionality. Compensation for clock skew must be performed on all source drivers, taking into account the total capacitance of all daughter boards that may be present within the system, as detailed in Chapter 4 and calculated using Eqs. (4.9) through (4.12).

7.4 INTERCONNECTS

Concerns exist for interconnects used on both backplanes and daughter cards. This is especially true when a large number of connectors are provided. When many connector slots are provided in an assembly, there exists a sum of intrinsic device delays that occur between loads connected to the bus. This is in addition to the intrinsic line delay of the boards that are inserted into the backplane. Signal degradation or performance with these delays should be evaluated. Our concern is with the capabilities of the total assembly related to I/O data transfer. I/O data transfer includes the source driver placing a signal on the backplane or daughter card, and then sending this signal to a load located elsewhere on the bus.

A typical backplane I/O connector must be chosen that is capable of handling the edge rates without signal degradation. To maintain signal integrity, the impedance mismatch at the connector boundary must be minimal. Sufficient ground pins and power pins must be provided throughout the connector to maintain a constant impedance match, as shown in Fig. 7.1. An impedance mismatch will cause common-mode RF currents and alter the EMI spectral profile simultaneously. This altered profile will be coupled to adjacent signal pins, and it can intermodulate differential-mode RF currents between signal and ground pins. All of these effects can also cause possible noncompliance with EMC regulations.

A backplane or daughter card connector used for high-speed applications must take into consideration the impact on signal transmission quality. Connectors, like signal traces and components, contain inductance, capacitance, and resistance. Parasitics from clock harmonics and common-mode RF currents will also be present, usually at those frequencies of greatest concern.

Design techniques for I/O connectors include the following recommendations:

1. Keep all discontinuities as short as possible so that the propagation intervals are a small percentage of the edge times.
2. Use as many ground connections as possible within the allocated space or pinout. Take into consideration the aspect ratio of the board and maximize the number of ground connections.
3. Establish a common ground within the connector.
4. Use appropriate dielectric constant board materials.
5. Maintain the ground path length as close as possible to the signal path between both signal and ground.

7.5 MECHANICAL

If possible, provide additional real estate on the backplane for any required filtering and termination of both internal and external I/O connectors and backplane power. High self-resonant frequency bypass capacitors for I/O cables and interconnects, if required, are installed in this additional area. If cables are attached to the backplane, and ground plane image references are to be connected to the logic grounds of the backplane, cable shielding overshields (braids and/or foil) are to be connected to the chassis-card cage ground. For this reason, the connector locations of external cables with overshields should be near the perimeter of the backplane to facilitate overshield connection to chassis.

All return and shield planes within a motherboard are generally RF bonded at every ground stitch location. DC power and return connector pins should also be equally spaced within the connector to minimize ground loops that may be created by poor aspect ratios, ground plane discontinuities, or poor trace routing. Route traces between different layers that are mutually adjacent and not separated by a ground plane at a 90° angle (horizontal versus vertical routing) to prevent crosstalk coupling between planes, thus enhancing signal integrity. Conversely, *do not* route traces at a 90° angle *on the same routing plane.*

Isolation or separation between traces internal to a backplane, daughter card, or motherboard is achieved through use of either the 3-W rule, by adding guard traces on both sides of a critical trace, or adding shunt traces between stacked routing layers.

7.6 SIGNAL ROUTING

Avoid vias between planes for all traces—clocks and signals. Each via adds approximately 1 to 3 nH of lumped inductance to the trace. Use of many vias may also make traces susceptible to poor signal quality and degraded EMI performance. If possible, route critical traces on the same signal plane. Daisy-chaining traces is permitted *only* if the load components are located adjacent to each other and/or in intervals that are compatible with the reflection and timing diagram. In all other cases, use terminated radial trace routing, identical to clock traces described in Chapter 4. Provide appropriate termination to the traces, with a minimal number of fanouts per driver. (A radial series termination scheme maintains functional signal quality as well as minimizing EMI noise by removing overshoot, undershoot, and ringing on signal traces as detailed in Fig. 4.11.)

When using I/O connectors and interconnects, minimize short stubs, also known as "T-stubs," that sometimes occur during routing. Keep this T-stub to a length of very short propagational time when compared to the edge time speed, even on non-clock or non-periodic signal traces. T-stubs always should be avoided on periodic signal or clock lines. T-stub lengths can be critical with I/O devices such as SCSI interconnects.

If a T-stub must be used because of problems with layout or routing, it must be as short as possible. Relocate components to remove T-stubs created by the CAD autorouter. There are applications where "Tee" signal routing can be made to function correctly. This method can be accomplished by feeding the center drive leg of the signal to two identical, and propagationally short, lengths of the "Tee" arms. The lengths of the two arms must be identical and short terminations may be placed at the end of each arm, if required. Use the measurement feature of the CAD system to determine this length. If necessary, serpentine route the shorter trace until it equals its counter-trace length exactly.

A potential or fatal drawback of using "Tees" lies in future changes to the PCB. If a design engineer or CAD person who has no knowledge of this "Tee symmetry" requirement makes a change to the layout, routing, or driver device edge times to implement rework or redesign, inappropriate changes to the trace may occur inadvertently, creating EMI and/or functionality problems.

7.7 TRACE LENGTH/SIGNAL TERMINATION

For standard-speed TTL logic, trace termination may not be required if the signal length is short in comparison to edge time. For higher-speed components, backplane terminators are usually necessary. The bus driver (e.g., 74xxx244) on one daughter board must be designed for driving terminated loads to other daughter boards through all connectors located on the backplane. Make the parallel terminator, if used, the last item on the signal trace route, with the driver circuit the first item on the bus. If using multiple drivers from different slots in the design, use high-current drivers in place of high-speed drivers to match the loaded bus impedance. Furthermore, due to the possibility of obtaining a distorted waveform, use the slowest possible receivers (functionally appropriate), properly balanced and terminated.

All other concerns and discussions related to trace length and signal termination are presented in Chapter 4.

7.8 CROSSTALK

Crosstalk in backplanes is a major concern when many traces are routed in parallel and are generally spaced close together. Signal routes may be very short to extremely long (i.e., electrically, with respect to propagation delay and signal edge rate). With long routes, proper termination is required to remove transmission line problems generated between source and load while enhancing signal quality. Detailed discussion on crosstalk is presented in Chapter 4.

If a very large backplane is needed, and the traces are electrically long, use of transmission line theory is required. Differential-mode paired signals may be enhanced for crosstalk rejection by placing the two traces in parallel, close together, and then enforcing the 3-W rule outside of the boundary formed by the pairs. This technique is illustrated in Fig. 7.4

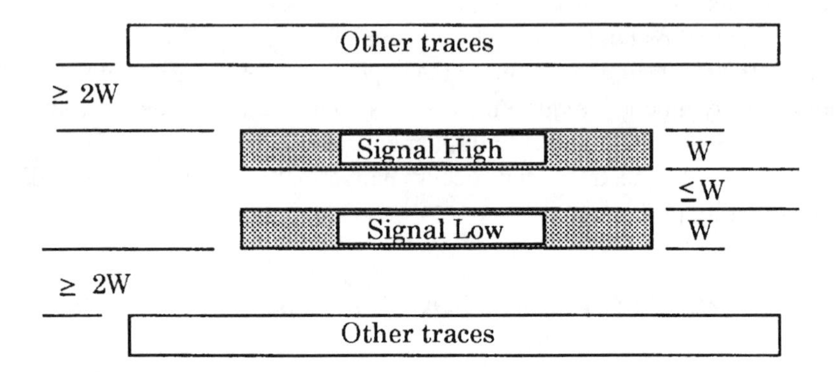

Fig. 7.4 Differential pair routing

The easiest and best design technique to prevent or minimize crosstalk is trace separation. Techniques include use of the 3-W rule when traces are routed on the same plane, or orthogonal (90°) when routed on adjacent signal planes (horizontal vs. vertical routing). Another technique to use for crosstalk control on the same routing plane is to separate parallel traces at 2 mils/inch (0.002"/in.) of trace length.

7.9 GROUND LOOP CONTROL

If a one-, two-, or four-layer backplane is used, proper attention must be made to minimize ground loops between power and ground traces. Route signal lines with as many ground traces interspersed as possible, interconnected to the main ground through the interface connector on the adapter card. Always route power and ground adjacent to each other, as shown in Fig. 7.5, if power and ground planes are not provided.

As seen in Fig. 7.5, traces are routed in stripline, or parallel trace (for two layers), fashion—signal trace adjacent to a signal return (ground) trace. The main power and ground traces are routed in the middle of the

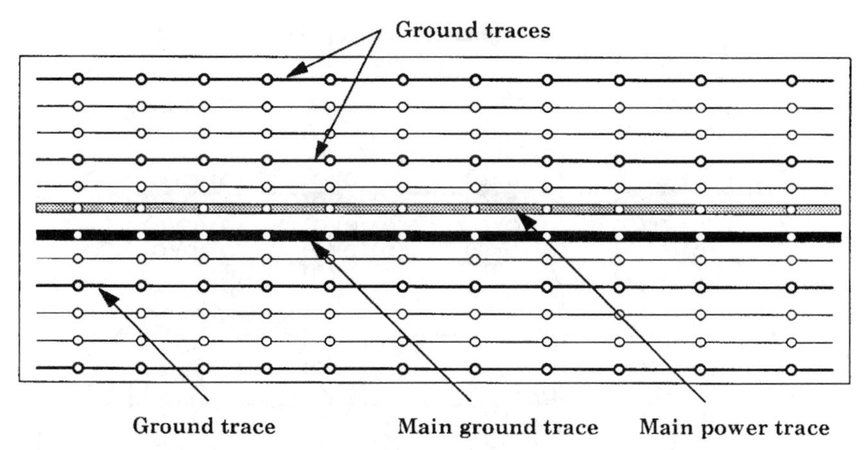

Note that the main power and ground trace (from power supply) is located in the middle of the board. This is to maintain better uniformity in power distribution throughout the entire assembly.

Fig. 7.5 Ground loop control in a backplane

board for better uniformity of power distribution. Also, routing the main power and ground traces adjacent to each other minimizes ground loops in the power section of the backplane. Use of multiple signal return (ground) traces in a backplane minimizes crosstalk between traces and allows separation of high-speed, high-threat signal traces (sources) from other sensitive and corruptible traces (victims), such as reset, alarm, and so forth.

7.10 GROUND SLOTS IN BACKPLANES

A common practice when designing backplanes is to select a connector with pressfit (stake) or through-hole pins. Newer-technology connectors with high pin counts are available in surface mount form. Surface mount connectors route signal traces to inner layers through vias. Depending upon the connector method chosen, ground slots may be accidentally designed into the backplane assembly.

A ground slot is a continuous discontinuity along the length of the connector. This is shown in Fig. 7.6. A ground slot is created because clearance holes for the pins of the connector are larger than they need to be. These clearance holes overlap and create a continuous discontinuity in the plane. There must be sufficient distance between pins for routing traces. Adjacent to each routing layer in the stackup is a power or ground plane used for not only power connection to the mated boards, but also for signal return of RF currents (image plane). With large overlapping cutouts

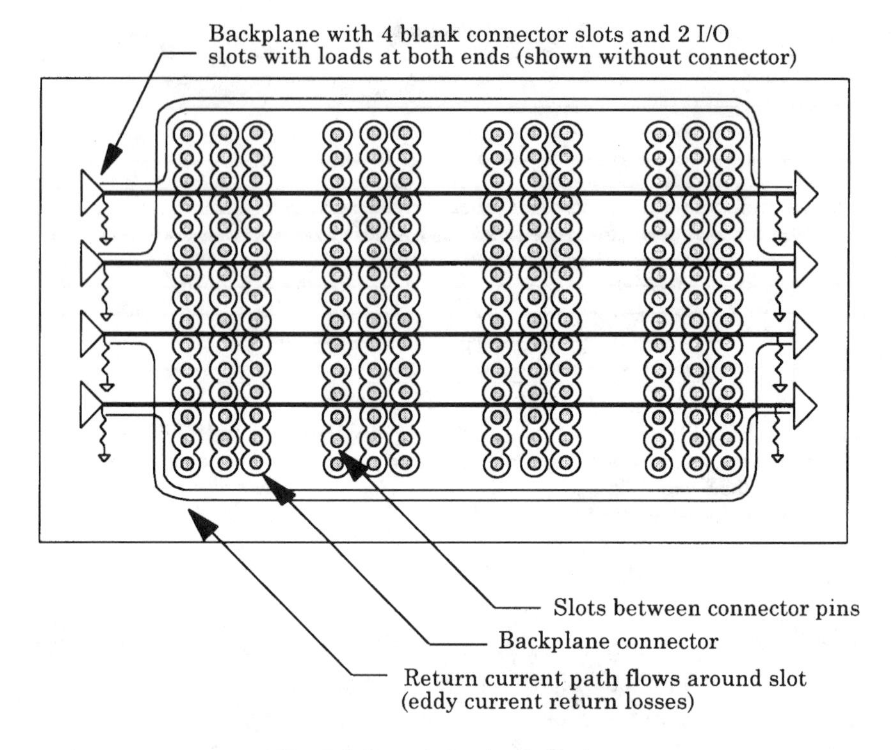

Fig. 7.6 Ground slots in a backplane

on all layers of the backplane, these solid planes will have discontinuities that prevent RF currents from returning back to their source in a straight-line manner. Under this situation, crosstalk may occur due to the mutual inductance that exists between the traces sharing a common signal return path. In addition, inductance added to the trace may create unwanted common-mode currents while degrading the edge rate of clock signals propagated between load boards at opposite ends of the backplane. Since the return current eddys around the slots, causing a phase shift in the return current, flux cancellation in the transmission line is also skewed, affecting signal quality as well as EMI.

As observed in Fig. 7.6, return currents between daughter card slots in the backplane cannot flow directly under their respective trace route. Instead, the return current is diverted around the ends of the connector due to the continuous ground slot created by oversized holes. To remedy this problem, ensure that ground clear-outs around each connector pin have ground continuity. The diverted current also increases the inductance of the trace. An increase in inductance slows the rise time of signals routed between source and load boards.

It is also seen in Fig. 7.6, that return current from one trace is parallel (actually overlaps) to the return current on an adjacent trace. This overlap (not shown in the figure) allows a large mutual inductance between traces to occur for return currents sharing the same return plane.

The effective inductance of this trace is calculated as follows:

$$L = 5d \ln\left(\frac{d}{w}\right) \tag{7.1}$$

where

L = inductance, nH

w = trace width, in.

d = slot length of connector, inches (extent of current diversion away from the signal trace)

Inductance is not related to the width of the ground slot. Inductance is related only to the perpendicular length of the slot. Any slot length will cause current diversion. Since current division is based only on slot length, traces closest to the edge of the connector will have less current diversion than traces routed in the middle of the connector (longer return current trace length).

8

Additional Design Techniques

8.1 TRACE ROUTING FOR CORNERS

Attention to detail must be observed when designing with subnanosecond transitions to avoid a discontinuity that occurs during normal routing of signals. When a trace makes a sharp bend on the board, its capacitance per unit length will increase, while its inductance per unit length will decrease. This is shown in Fig. 8.1, and is true for sharp angles of 90° or more. Right-angle bends look like a capacitive load attached to the transmission line. A capacitive load will round off or change the edge rate of a signal trace. To prevent capacitive loading, do not incorporate right-angle corners. This is especially critical when clock signal edges are 2 ns or faster. Almost all CAD programs allow prevention of routing traces at 90° angles. Do not turn off or disable this routing feature. Since clock signals must always be manually routed, guarantee that these traces are first routed without 90° angles before routing the rest of the board.

Rounding the outside corner of the bend leaves a constant width dimension that is physically smaller than a 90° or 45° angle. With 90° angles, an increase in trace width exists that contributes extra unwanted parasitic capacitance. An edge that is less than 90° reduces the amount of reflection and signal rise degradation for signals rounding the corner. When the corner is chamfered (45° angle), a reduction of up to 57 percent of the capacitance is achieved. Chamfered corners work up to 10 GHz. Above 10 GHz, round corners perform better. This effect is partially shown in Fig. 8.1.

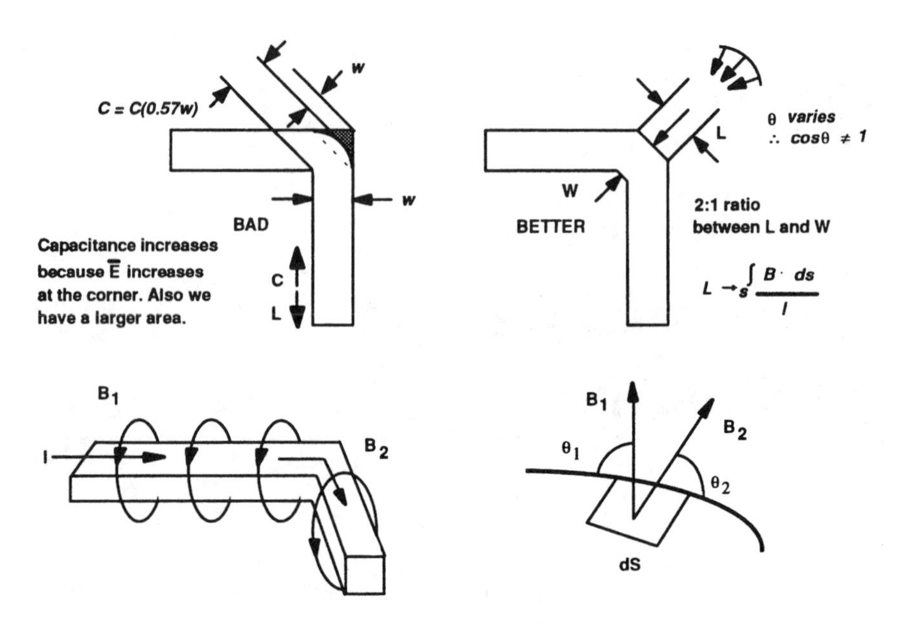

Fig. 8.1 Trace routing for corners (Source: CKC Laboratories, reprinted by permission)

The technical reason of why 90° angles are poor for EMI is that inductance is derived from $L \rightarrow \int B \bullet ds$, as shown in the bottom two details of Fig. 8.1.

If $\theta = 90°$, then $\sin\theta = 1$. If $\theta < 90°$, then $\sin\theta < 1$ and

$$B \bullet ds \text{ for } 0 < \theta < 90$$

$$<$$

$$B \bullet ds \text{ for } \theta = 90$$

where **B** = magnetic field flux density.

Another way to explain this design constraint of not routing traces at 90° angles is presented with a simple analogy. This analogy is used to describe the right-hand rule for magnetic flux and current flow. For example, RF current likes smooth corners, like an automobile. If you ask RF current, or your automobile, to make a sharp 90° turn, lines of magnetic flux, or your car, will have trouble negotiating this sharp 90° turn, thus causing potential signal nonfunctionality. In the case of RF, magnetic lines of flux will vary in a nonuniform manner because the angle of flux linkage must change in an abrupt manner.

When performing artwork cleanup of a PCB for all traces prior to release (usually called "glossing the board"), make sure that all trace corners that were routed at 90° angles are converted to 45° angles.

8.2 HOW TO SELECT A FERRITE DEVICE

It is a well known fact that ferrite devices (bead-on-leads, toroids, cores, split cores, wound beads, and so on) attenuate RF energy on traces. The use of ferrite devices is mentioned throughout this design guide. The biggest difficulty in using a ferrite component is selecting the proper device for a specific application. Usually, a trial-and-error method is employed by the EMI or design engineer during attempts to solve a radiated or conducted emission problem. In reality, selection of a ferrite device is quite simple [1].*

There are three common ways to select ferrites for suppression of unwanted signals. Their relative use in real-world circuits depends on the particular application.

1. Use a ferrite as a shield to isolate a conductor, component, or circuit from stray electromagnetic fields.
2. When a ferrite device is used with a capacitor, a lowpass filter is created that is inductive-capacitive (LC) at low frequencies and dissipative at higher frequencies.
3. Use a ferrite device to prevent parasitic oscillation or to attenuate unwanted signal coupling along component leads, interconnecting wires, traces, or cables (used as a lossy element).

The selection of a particular ferrite material is based on the impedance that the device presents to the circuit. This impedance is based on the permeability, μ, of the material. The impedance of the ferrite material in reality is a series combination of inductive reactance ($j\omega L$) and loss resistance (R), both of which are frequency dependent ($R + j\omega L$). The real component represents the loss portion, and the imaginary component represents the reactive.

At lower frequencies, the impedance is primarily inductive reactance, which is a function of the material's permeability. Most unwanted signals are thus reflected. At higher frequencies, the inductive reactance decreases, causing the total impedance to increase; thus, unwanted signals are absorbed.

Examining the real (reactive) component, it is observed that the μ of different materials ranges from 10 to 15,000. This is principally due to the ferromagnetic resonance of the material. The higher the permeability, the lower the resonant frequency.

* The discussion of ferrite materials is reprinted courtesy of Fair-Rite Corporation.

When selecting a ferrite material, one must first know the frequency or range of frequencies to be suppressed, and those frequencies that must be passed. Different ferrite families (i.e., materials used in manufacturing) have different permeability, inductive reactance, and loss resistance. The most commonly used material and their filtering ranges (based on permeability) are shown in Table 8.1.

Table 8.1 Frequency Range of Ferrite Materials

Permeability (μ)	*Frequencies Suppressed*
2500	30 MHz or below
850	25 to 250 MHz
125	200 MHz and above

Custom materials with different permeability values are available from manufacturers of ferrites for use at a particular frequency or range of frequencies. Generally, the higher the permeability, the lower the optimum attenuation frequency. Conversely, the lower the permeability, the higher the attenuation frequency. This is because low-frequency attenuation is reflective and high-frequency attenuation is limited by the core and circuit resonance.

In addition to selecting a material for a certain range of frequencies, consider the attenuation desired for a particular application. This attenuation is calculated using Eq. (8.1).

$$\text{Attenuation} = 20\log_{10}\left(\frac{Z_s+Z_{sc}+Z_L}{Z_s+Z_L}\right)\ \text{dB} \tag{8.1}$$

where:

Z_s = source impedance

Z_{sc} = suppressor core impedance

Z_L = load impedance

Equation (8.1) is dependent on the impedance of the source generating the noise and the impedance of the load receiving it. The result of this equation is generally in complex form and difficult to solve.

Selection of a ferrite material is not based on the permeability value alone. Consideration of the core size, the environment, biases, and resistivity must be taken into account.

The core size or shape determines the maximum impedance of the core for a particular package size. Generally, as the length of the core increases, impedance increases versus the diameter for the same volume. To achieve greater impedance, select a core with greater length, either circular, flat, or toroidal. Table 8-2 summarizes three different core sizes and the impedances presented to the circuit.

Table 8.2 Comparison of Impedances of Common Ferrite Bead Materials

Bead Type and Dimensions ($L \times OD \times ID$ in mm)	*Impedance in Ω*					
	1 MHz	*5 MHz*	*10 MHz*	*20 MHz*	*30 MHz*	*50 MHz*
μ = 850, 3.25 × 3.5 × 1.6	2	8	13	20	28	32
μ = 2500, 3.25 × 3.5 × 1.6	11	26	32	37	37	35
μ = 850, 7.5 × 7.65 × 2.25	5	18	29	40	58	61
μ = 2500, 7.5 × 7.65 × 2.25	25	47	58	61	61	60
μ = 850, 11.1 × 5.1 × 1.5	14	41	66	95	110	115
μ = 2500, 11.1 × 5.1 × 1.5	46	100	125	160	160	155

Source: Fair-Rite Corporation, reprinted by permission.

The *environment* will affect the magnetic parameters that change with temperature and field strengths. An increase in temperature will cause a decrease in the overall impedance. For use at elevated temperatures, select a ferrite material that decreases at a slower rate of change per degree C.

Bias refers to the amount of dc current passed through the ferrite core. An increase in bias will decrease the impedance of the core more than any other parameter. The magnetic field strength can also cause significant degradation in impedance in the lower frequency ranges. To increase the dc-carrying capabilities of a device, select a core shape with a built-in air gap. The larger the gap, the less effect bias will have on the impedance.

The *resistivity* of ferrite material varies, depending upon the value of dc or ac current passed through the bead. This resistivity may cause excessive attenuation of a "wanted" signal, as if the ferrite bead were replaced by a resistor.

To increase the impedance of shield beads or cores, additional turns of wire may be added. The impedance will increase in direct proportion to the turns squared. However, the frequency at which maximum impedance is reached is lowered due to the additional capacitive effects of the wire. The net effect is to narrow the effective range of frequencies available for suppression of unwanted energy.

8.3 GROUNDED HEATSINKS

Grounded heatsinks represent a new concept in PCB suppression that finds use in specific applications and only for certain components. Grounded heatsinks are sometimes required when using VLSI processors with internal clocks in the range of 75 MHz and above. These CPU and VLSI components require more extensive high-frequency decoupling and grounding than do most other parts of a PCB.*

New technology in wafer fabrication allows component densities to easily exceed one million transistors per die. As a result, many components consume 15 watts and more of power. Certain reduced instruction set computing (RISC) CPUs consume 18 to 25 watts of power and require separate cooling provided by a fan built into a heatsink, or located adjacent to a fan or cooling device, and so on. Since these high-power, high-speed processors are being implemented in more designs, special design techniques are now required for EMI suppression at the PCB level.

Examining the function of a heatsink in the thermodynamic domain, we note that removal of heat generated internal to the processor must occur. Components that dissipate large amounts of heat are usually encapsulated in a ceramic case, since ceramic packaging will dissipate more heat than will a plastic package. Ceramic cases also cost more. Certain components, due to large junction temperatures between gates and gate quantity, generate more heat than the ceramic package can dissipate; hence, a heatsink is required for thermal cooling.

Having briefly discussed the function of heatsinks in the thermodynamic domain, the metal heatsink is now examined in the RF domain. For proper thermal implementation and use of heatsinks, a thermal insulator (silicon compound or mica insulation) is provided. These compounds are generally electrically nonconductive; however they contain excellent thermal properties for transferring heat from the component to the heatsink. Examining metal heatsinks in the RF domain, the following characteristics are observed, as illustrated in Figures 8.2 and 8.3.

- Wafer dies operating at high clock speeds, generally 75 MHz and higher, generate large amounts of common-mode RF current within the package.
- Decoupling capacitors remove differential-mode RF current that exists between the power and ground planes and signal pins.

* The grounded heatsink model was first identified and conceptualized by W. Michael King, and placed into commercial use by Mark I. Montrose for use with the R4000 RISC CPU processor family from MIPS Computer Systems, Inc.

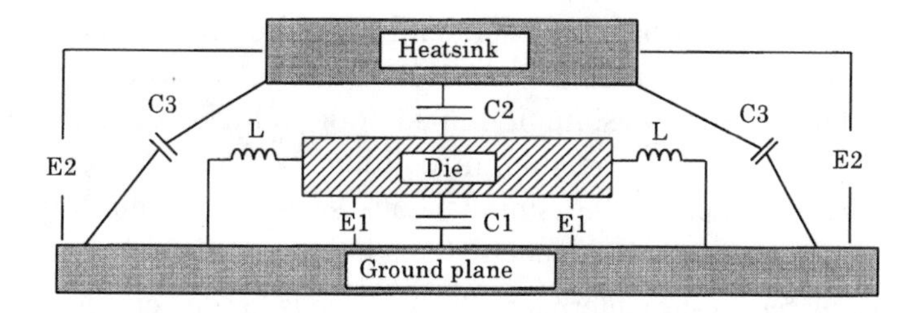

L = Package lead inductance
C1 = Distributed capacitance from the die to the ground plane
C2 = Distributed capacitance from the heatsink to the die
C3 = Distributed capacitance from heatsink to ground plane or chassis
E1 = Potentials developed between die and logic ground plane
E2 = Potentials coupled to heatsink, resulting in an EMI antenna

Typical self-resonant frequency of VLSI processors is approximately 400-800 MHz

Fig. 8.2 Electrical representation of a heatsink

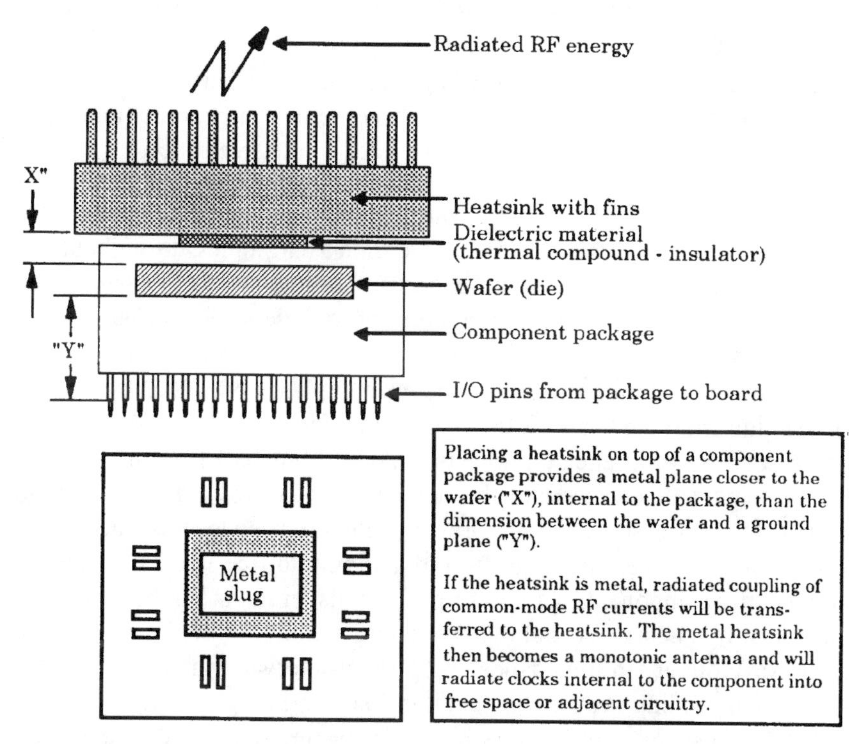

Fig. 8.3 Ungrounded (normal) heatsink mechanical representation

- Certain ceramic packages contain pads on top of the package case to provide additional differential-mode power filtering required by the large power consumption in addition to high-frequency decoupling. These capacitors minimize ground bounce and ground-noise voltage created by the simultaneous switching of all component pins under maximum capacitive load.
- The wafer (or die) internal to the package is located closer to the top of the device (dimension "X") than the bottom of the package (dimension "Y"). Therefore, height separation from the die to an image plane internal to the PCB is greater than the height of the die to the top of the package case. Common-mode RF currents generated within the wafer have no place to couple; hence, RF energy is radiated into free space. Differential-mode decoupling capacitors will not remove common-mode noise created within the component.
- Placing a metal heatsink on top of the component provides an image plane closer to the wafer than the image plane on the PCB; thus, tighter *common-mode* RF coupling occurs between the die and heatsink than between the die and the first image plane of the PCB.
- Common-mode coupling to the heatsink now causes this thermodynamically required part to become a "monotonic antenna," which is perfect for radiating RF energy into free space.

The net result of using a metal heatsink is the same as placing a monotonic antenna inside the unit to radiate clock harmonics throughout the entire frequency spectrum. To de-energize this antenna, the heatsink must be grounded. This is a very simple concept to understand, yet is virtually ignored within the field of EMC and PCB design for suppression.

Heatsinks must be grounded to the ground planes of a PCB by a metal connection from the heatsink to the ground planes on all four sides of the processor. Use of a fence (similar to a vertical bus bar) from the PCB to the heatsink will encapsulate the device. This creates a Faraday shield around the processor and thus prevents common-mode noise generated within the component package from radiating into free space or coupling onto nearby components, cables, or peripherals, or radiating through aperture slots. It is necessary that the grounding method for the heatsink be of microwave quality in impedance and ground-interval dimension! A technique for providing heatsink grounding is shown in Fig. 8.4.

RISC processors and VLSI components generally have high self-resonant frequencies. This high self-resonant frequency is a combination of the manufacturing process and internal clock speed, in addition to the impedance present in the power planes during maximum power consumption. As

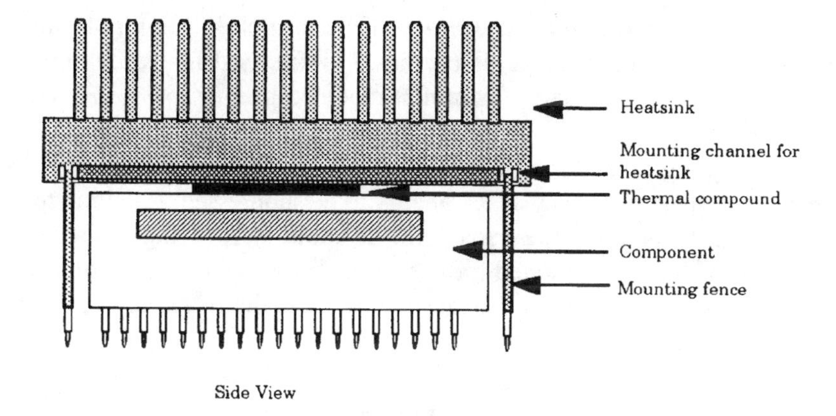

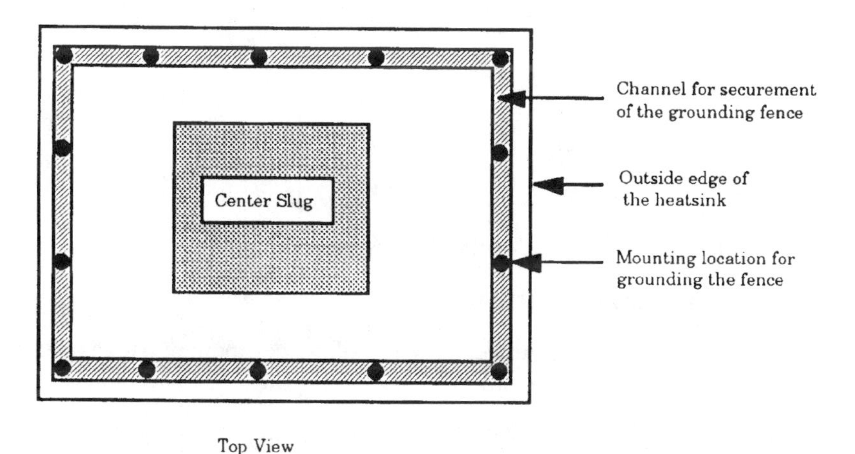

Fig. 8.4 Grounding the heatsink

a result, VLSI components can radiate more RF energy than most other components if internal RF suppression techniques are not incorporated by the component manufacturer. Any attempt to remove this self-resonant RF frequency using standard suppression techniques is almost impossible except through use of a heatsink as a *common-mode decoupling capacitor*. This heatsink is used in conjunction with differential-mode capacitors located on both the top of the ceramic package, in addition to decoupling capacitors located under the device directly on the PCB.

A grounded heatsink must always be at ground potential. The active component is at RF voltage potential. The thermal compound is a dielectric insulator between two large plates. The definition of a capacitor exists as described by Eq. (3.5). Thus, a grounded heatsink works as one large "*common-mode decoupling capacitor*," while discrete capacitors located on top of the device package or on the PCB are used for "differential-mode" decoupling. This large "common-mode" capacitor absorbs RF currents generated within the processor.*

A decoupling capacitor will only work if its self-resonant frequency is above the highest frequency of RF energy to be suppressed. A unique feature of this common-mode capacitor (in addition to being a Faraday shield) is its ability to be tuned for a specific frequency or application. To tune a heat sink for optimal resonance, determine the frequency of interest, X_c, and use Eq. (8.2).

$$X_c = \frac{1}{2\pi fC} = \frac{d}{2\pi f\varepsilon A} \tag{8.2}$$

where

$C = \varepsilon A/d$

X_c = capacitive reactance

f = desired resonant frequency of the component

ε = dielectric constant of the thermal compound

A = area of the parallel plates

d = distance spacing between the top of the component package and bottom of the heatsink

The height, d, is created by the thickness of the dielectric material (thermal compound). The manufacturer of the compound (or thermal pad) can provide the dielectric constant of the material. Since X_c, f, ε, and A are known, calculate d, the thickness of the thermal compound, by Eq. (8.3).

$$d = X_c(2\pi f\varepsilon A) \tag{8.3}$$

Using a grounded heat sink creates:

1. a thermal device to remove heat generated within the package
2. a Faraday shield to prevent RF energy created from the clock circuitry within the processor from radiating into free space
3. a "common-mode" decoupling capacitor that removes common-mode RF currents generated directly from the die or the wafer inside the package

---

* The use of a heat sink as a return plate for high-performance, high-frequency decoupling capacitance was first conceptualized by W. Michael King.

If a grounded heatsink is implemented, make sure that the grounding fingers of the fence (spring fingers or other PCB mounting method employed) are connected to all ground planes in the PCB on at least 1/4 inch centers around the processor. At each and every ground connection, install parallel decoupling capacitors, alternating between each ground pin of the fence with 0.1 μF in parallel with 0.001 μF *and* 0.01 μF in parallel with 100 pF. RF spectral distribution from RISC processors and similar components generally exceeds 1 GHz bandwidth. RISC and VLSI processors also require more extensive multipoint grounding around all four sides of the processor than do most other types of components. It is observed that these bandwidths complement the approximate $\lambda/4$ mechanical size of typical heatsinks, making them efficient radiators of EMI spectra.

8.4 LITHIUM BATTERY CIRCUITS

International safety agencies require protection against explosion from lithium batteries should an abnormal fault occur (short circuit or reverse bias during a charging cycle). Lithium battery circuits may consist of a discrete battery with passive components or be included as part of a nonvolatile RAM or nonvolatile clock calendar component. The battery must be provided with a reverse current protection circuit. This protection circuit consists of two diodes back to back or a diode and a resistor in series. A typical lithium battery circuit with protection components is shown in Fig. 8.5.

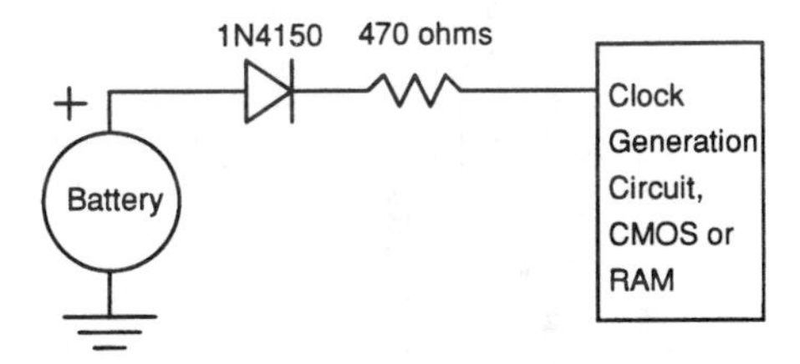

Fig. 8.5 Lithium battery circuit design

8.5 BNC CONNECTORS

BNC connectors require special design considerations. For I/O interconnects where the shield of the coax is used for RF shielding purposes, ground the shell of the BNC connector to *chassis ground* at the bulkhead connector or I/O panel through a low-inductance (low-impedance) path. Verify that this ground connection is to chassis ground and not to isolated or signal ground. *Never use a pigtail to ground the BNC connector shield to chassis ground.*

If, for reasons of desired isolation, the BNC connector is required to be "floating" from chassis ground, use an isolated BNC connector available from many vendors. These isolated BNC connectors guarantee that the shell and internal signal pin of the connector are isolated from chassis ground. For purposes of EMI containment and ESD rejection, a high-performance capacitor must then be connected between the BNC shell and the chassis. Some manufacturers of isolated BNC connectors include such capacitance within the isolated connector block (these are preferred), but additional protection devices may be required for ESD events. It is to be noted that this scheme will cause problems if not properly implemented. This is shown in Fig. 8.6.

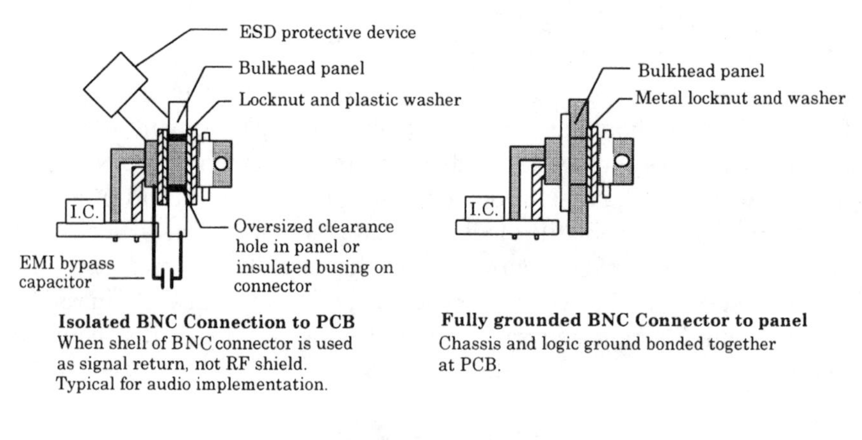

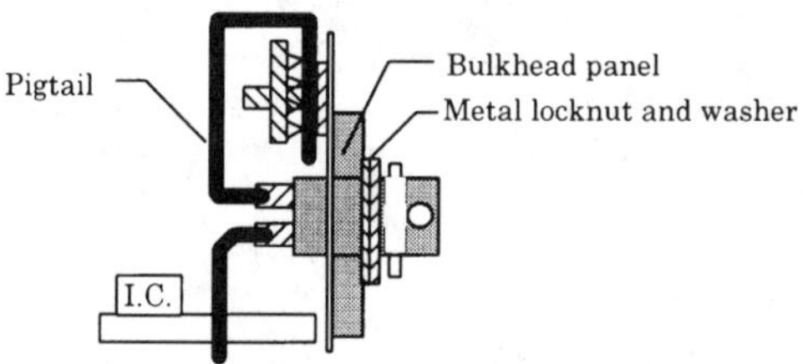

Fig. 8.6 BNC connectors

Place the source or I/O driver as close as possible to the BNC connector. Design the impedance of the signal trace (see Chapter 4) to match the impedance of the coax (transmission line), typically 50, 75, or 92 Ω.

Never use pigtails under any condition to connect the shell of the BNC connector to chassis ground or to any other ground system. Measurements are well documented showing a 40 to 50 dB difference between a pigtail and a 360° connection of the cable shield to the BNC connector shell in the 15 to 200 MHz region for RF emissions. In addition to improvement in reducing RF emissions, a greater level of ESD immunity is provided due to lower lead length inductance presented to the ESD event. For most applications, the recommendation for cable shields is to connect the cable shield to the BNC connector shell in a 360° fashion. This backshell then mates with a bulkhead panel containing a solid metallic contact with chassis ground [2].

Many excellent reference books discuss additional techniques for I/O interconnects using a large number of connector families and types.

8.6 FILM

Due to the increasing density of PCBs, the assembly process needs to be monitored closely by the designer to defend EMI signal integrity and EMI performance parameters that are signal-impedance dependent. The following features can be helpful:

1. *Test coupons.* Test coupons are traces that are routed on an external part of the main PCB and connected to test points at each end. These test traces allow for easy measurement of the impedance of the internal planes in the board. This trace demonstrates quality control of fabrication by a vendor by ensuring that the board was designed with a desired impedance. These test traces and pads are usually found on breakaway islands and not on the main artwork or final assembly. For backplanes and very high-speed PCBs, these test traces should be placed directly on the artwork, if not actually part of the design.
2. *Layer stackup window.* Etched onto each layer of a PCB is a number (in a box) that reflects the existence of a layer internal to the board. The advantage of using a layer stackup window is to help identify how many layers physically exist. If experimentation on routing layers and image planes occurs for performance optimization (during debug and EMI evaluation), this window will assist in identifying different board stackup schemes and the number of layers used.
3. *Test points.* Test points are used to assist the design engineer in analyzing the signal integrity of critical nets.

Prior to the generation of the physical stackup dimensions (spacing between layers), consult a PCB manufacturer for optimal design parameters if details on how a board is manufactured are not well known or understood. These vendors can provide detailed information on dielectric material, trace width dimensions (for impedance control), height of each signal layer to a plane, and so forth. A typical stackup is shown in Fig. 8.7 for a 10-layer board. Stackup assignment is discussed in Chapter 2.

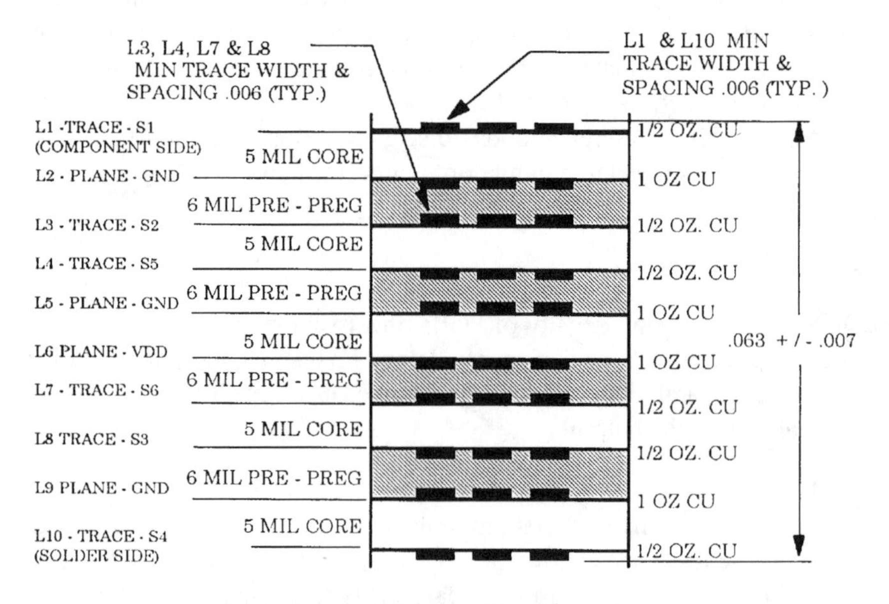

Fig. 8.7 Typical 10-layer PCB stackup scheme

Core and prepreg in the figure refer to material usage. The core material is the bare fiberglass (or similar) material that contains a thin sheet of copper laminate on both sides. Prepreg is the glue material (epoxy) placed between each pair of cores. Prepreg is heated and pressed into place during production. The thickness of the prepreg layer determines the spacing between the core layers. The prepreg material must have the same dielectric constant as the core material to maintain proper impedance control between planes. Core and prepreg alternate in multilayer boards.

Consider a two-layer board as a substructure of a multilayer board. The outer two layers are the copper laminate sheets. The material between these two copper sheets is known as the core material. Figure 8.7 shows a typical 10-layer board with different thickness of copper planes, core, and prepreg material. Note that, since S6 on layer 7 is directly adjacent to the VDD plane of layer 6, it is more susceptible to noise injection.

It is observed that all signal layers are 1/2 oz copper, while the power planes (voltage and ground) are 1 oz copper. For functionality of operation, 1 oz copper, or twice the thickness of the signal planes, is generally required. This is because power planes must provide high levels of voltage and return currents to all components. To prevent overheating of the power planes because of a large voltage (IR) drop internal to the plane, caused by power consumption of all components, an additional 1/2 oz (or 2x) copper is required.

In a multilayer board, start by defining the core. These inner layers are generally the power and ground planes. If designated as signal planes, etching is performed to define the trace structure. The thickness between the opposing layers and the core depends on the thickness of the original laminate. If the overall height dimension of the board is 0.063 inches, the thickness of the core for a four-layer stackup will be greater than the spacing for a higher-density board. This is because, with more layers, smaller spacing between layers must occur to maintain the same overall height dimension. With smaller distances, the trace width must be altered to maintain proper impedance, as discussed in Chapter 4. An example of height distance variation for various stackup schemes is shown in Fig. 8.8.

In examining Fig. 8.8, trace signals (not shown on any particular layer) are laminated on a copper plane. This lamination layer is then located adjacent to a prepreg layer. Since the prepreg is a glue, the overall height thickness of the trace will not increase the distance spacing, given that the trace thickness will be absorbed in the prepreg material. If different thickness layers are used, adjust the distance spacing appropriately. Impedance control will also alter these numbers. Figure 8.8 is representative of a typical stackup assignment that would be provided by the design engineer to the board vendor for fabrication.

For high-speed applications, locate the power and ground plane as close together as possible. This creates an internal decoupling capacitor that is useful for removing RF current created by components and traces. This close spacing will also reduce power supply noise on the power planes.

If additional planes are required, make these planes at ground potential. It is common to have one power plane and multiple (two, three, or more) ground planes. Multiple power planes will allow return currents existing in the image plane to jump between the power and ground planes, creating a voltage gradient between these planes and RF current and the possibility of return flux phase shifts. These conditions must be avoided at all times.

Connect the multiple ground planes together in as many places as possible with ground vias if component ground pins are not located in close proximity throughout the entire board.

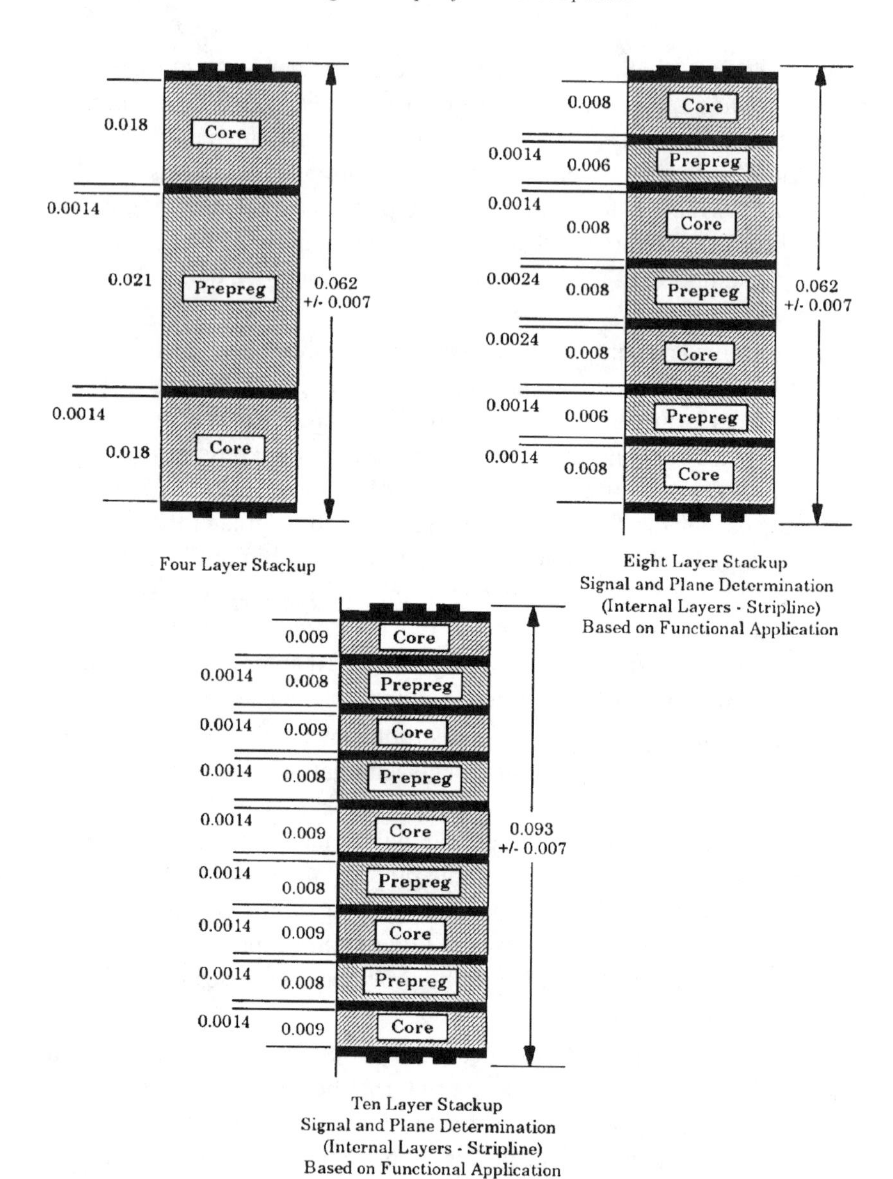

Note: The above dimensions assume 1 oz. thick copper planes (0.0014). Thickness of 2 oz copper is 0.0028.

If functional requirements mandate 1/2 oz copper, the thickness of these particular planes will be 0.0007.

With smaller thickness planes, the distance spacing between planes (prepreg) must be larger to maintain overall height of the board.

If the board must be impedance controlled, the dimensions shown will also be different based upon functional application and required impedance.

Fig. 8.8 Stackup details

Do not place solder mask on any ground stitch connection (to chassis ground) or on any localized ground plane. This includes ground points (chassis ground screws, gaskets), I/O bulkhead connectors, I/O adapter bracket mounting holes, ESD guard bands, and clock circuit pads on both the component (top) and solder (bottom) layers. Solder mask changes the dielectric constant of the board material. This dielectric constant could be an important design consideration in certain applications. Solder masks also prevent RF bonding between critical ground connections and mounting securement hardware.

Remove all unconnected vias from the artwork prior to production of the film. An unconnected via is defined as a pad on a plated through-hole that has no physical connection to any trace on any signal layer. This must be performed by the PCB designer during the final layout process. When a clock guard trace has been pulled away from a clock signal by a via, pin escape, or through-hole device, make sure there is no connection to this via on that layer (connection to this via must be made on a different layer).

For compliance with product safety standards, use of flame retardant material for the fiberglass assembly material is required. A rating of V-1 or V-0 is the minimum required flammability rating accepted by international safety agencies. The vendor's logo, date code, and flammability rating must appear on the bare board, on either the top or bottom layer. This labeling requirement applies to the board fabricator only. A notation on the assembly drawing must indicate this requirement.

8.7 REFERENCES

1. Parker, C.U. 1992. Choosing a Ferrite for the Suppression of EMI. Fair-Rite Products, Inc.
2. 1990. *Design Guidelines for PCB Layout*. CKC Laboratories.

Appendix A

Summary of Design Techniques

This appendix provides a checklist for various design procedures. Related sections of this book are indicated in parentheses. Information presented is in rules-driven format for easy implementation.

A.1 PRINTED CIRCUIT BOARD BASICS

1. Use proper topology for intended application; i.e., microstrip and/or stripline.
2. Use optimal stackup, depending on application and number of nets to route. Locate all signal routing planes adjacent to a solid image plane (2.1).
3. On two-layer boards, route power and ground traces radially (2.1).
4. Locate components based on radial migration from highest to lowest bandwidth (2.1).
5. Make power planes physically smaller than the ground planes per the 20H rule (2.2).
6. Select proper type of grounding assignment, product dependent: series, parallel, single-point, multipoint (2.3).
7. Use single-point grounding for low-frequency applications (audio, analog instrumentation, 60 Hz power systems, etc.) with clock rates of 1.0 MHz and lower. Use multipoint grounding for high-frequency systems using clock rates greater than 1.0 MHz (2.3).
8. Minimize RF ground loops between (2.4):
   - High RF energy level circuits and system ground
   - Functional subsections

- Multipoint ground locations
- I/O interconnects and associated control circuitry
- Power supply input terminals and system ground
- Card edge connectors and main system ground
- Opposite edges of the printed circuit board
- Cable shields and chassis ground

9. Calculate aspect ratio of all ground point locations and the straight line distance created by $\lambda/20$ of the highest frequency generated on the board. Place a ground stitch in the middle of this straight line path between these two ground points (2.4).
10. Always locate a signal routing plane adjacent to a solid image plane (2.5).
11. Never violate an image plane with a signal trace. Moats are acceptable in an image plane, provided the adjacent signal routing layer does not have traces cross the moated area. Be aware of ground plane discontinuities (2.5).
12. Never place three or more routing layers adjacent to each other. Each routing layer *must be adjacent* to a solid image plane (2.5).
13. Partition the printed circuit board into functional subsections. Separate high-bandwidth areas from medium- and low-bandwidth areas. Isolate each section using partitions or moats, if required (2.6).
14. Ground each partition and functional subsection to chassis ground in as many locations as possible to minimize ground loops (2.6).
15. Choose an appropriate logic family for functionality purposes. Do not use high-speed components when another logic family with a slower edge rate is acceptable. Components with fast edge rates generates greater amounts of spectral bandwidth of RF energy than devices with a slightly slower edge rate. Manufacturers of components rarely specify the minimum edge rate—only the maximum rise and fall time to guarantee functional performance. If in doubt, measure the actual edge rate and make component selection based on test results (2.7).
16. Minimize inductive trace lengths of components by not using sockets for through-hole devices (2.7).
17. Select logic components with power and ground pins located in the middle and not on the corners. This minimizes lead length inductance and ground loops created by decoupling capacitors (2.7).

18. Be aware of the peak inrush surge current into the device pins. This surge current may inject high-frequency switching noise into the power planes (2.7).

A.2 BYPASSING AND DECOUPLING

1. Select a capacitor based on intended use; decoupling, bypass, or bulk.
2. Bypassing and decoupling are functions of circuit resonance. Determine if the circuit is a series, parallel, or parallel C-series RL network. Calculate resonant frequency (3.1).
3. When selecting a decoupling capacitor, take into consideration the point source of charge a logic device requires for functionality. This is in addition to the resonant frequency required for removal of high-frequency RF currents developed from the component switching all pins simultaneously under maximum capacitive load (3.2).
4. Take into consideration the lead length of radial and axial capacitors when selecting a capacitor for a particular self-resonant range of frequency (3.2).
5. Capacitors decouple RF currents up to their point of self-resonance. Above self-resonance, the capacitor goes inductive and will cease to remove RF currents created by components. Certain logic families have a higher spectral distribution of RF energy than the self-resonant frequency of commonly used decoupling capacitors (3.2).
6. Provide decoupling capacitors on circuits that contain high-frequency RF energy or have clock edges faster than 3 ns. Calculate capacitance value for optimal performance and frequency range of interest. Do not guess this value or use the same capacitance value from previous designs (3.2).
7. Measure or calculate the self-resonant frequency of the printed circuit board's power and ground planes. These planes perform as a decoupling capacitor all by themselves. Use this built-in decoupling capacitor to maximum advantage (3.2).
8. Use parallel capacitors to bypass a larger bandwidth of RF spectral energy on *all* high-speed components and in high-RF-bandwidth areas (3.4).
9. When selecting parallel capacitors, remember that as the larger value capacitor goes inductive, the smaller value capacitor is still

capacitive. At a particular frequency, an LC circuit is developed between the two capacitors, and an infinite impedance could be generated with no decoupling provided at all. When this occurs, single-capacitor bypassing is generally sufficient (3.4).

10. Fewer decoupling capacitors may be better than many (3.4).
11. Power planes generally provide an adequate low self-resonant frequency decoupling for standard TTL components (3.5).
12. Keep lead lengths of capacitors as short as possible to minimize lead length inductance (3.6).
13. Place parallel bypass capacitors on all power and ground input connections on the printed circuit board in addition to components with edge rates faster than 3 ns (3.7).
14. Select components with power and ground pins located in the middle of the device and not on the corners (3.7).
15. Use sufficient quantity of bulk capacitors to provide a localized source charge of dc voltage and current. This is required for components that consume large amounts of power when all pins switch simultaneously under maximum capacitive load. Generally, a decoupling capacitor will perform the function of both bulk and RF current suppression (3.7).
16. Use bulk capacitors at all power connectors and at the opposite or far ends of the board. Also, locate bulk capacitors at the furthest location from the power entry connector. This is in addition to all components that consume large amounts of dc voltage and current. Bulk capacitors minimize dc voltage and current fluctuations (dropout or droop) that cause degradation of circuit functions (3.7).
17. Calculate the proper dc voltage rating for all capacitors (3.7).
18. If too many decoupling capacitors are used, excessive current draw from the power supply could occur, placing a strain on the power supply (3.7).

A.3 CLOCKS

1. Place clocks and oscillators in a separate clock generation area. Make provisions for a localized ground plane and doghouse (case shield) around the oscillator and related high speed, high current drivers. Locate clock generation circuits near a ground stitch location (4.1).

2. Always install clock circuits (oscillators, crystals, drivers, etc.) directly on the printed circuit board, not on sockets (4.1).
3. When using localized ground planes, observe the following (4.2):
   - Locate clock circuitry and localized ground plane next to an adjacent ground stitch location and bond the localized plane to this chassis ground connection.
   - In addition, connect the localized ground plane to the main ground plane using many vias.
   - Do not use solder mask on the localized ground plane.
   - Include support circuitry, drivers, buffers, and resistors in this localized ground plane area.
4. Maintain impedance control for all clock traces. Calculate impedance for both microstrip and stripline implementation (4.3).
5. Be aware of the propagation delay of signal traces routed either microstrip or stripline (4.4).
6. Calculate capacitive loading of all components and properly compensate with a series resistor and/or end termination (4.5).
7. The higher the switching speed (edge rate of the signal), the more important the series termination resistor from the clock driver must equal trace impedance Z_0. A perfect match exists when the impedance of the drive device source, Z_S, added to the value of the series termination resistor equals the trace impedance (4.5).
8. Decouple clock components with capacitors having a self-resonant frequency higher than the clock harmonics requiring suppression. Decoupling may include a single capacitor or two capacitors in parallel (4.6).
9. Printed circuit boards generally have a self-resonant frequency in the 200–400 MHz range. Use this built-in decoupling capacitor in the power planes to maximum advantage (4.6).
10. Minimize or prevent routing clock traces with vias. Vias add inductance to the trace (approximately 1–3 nH each). Vias could change the trace impedance causing possible non-functionality or EMI emissions (4.7).
11. The wider the trace, the less impedance presented to the circuit (4.7).
12. Do not locate clock signals near I/O areas. For traces within 2 inches of I/O, use the lowest-speed logic device possible. For traces within 3 inches of I/O circuits, use medium-speed logic.

This is not a requirement when functional partitioning is implemented (4.7).

13. Keep impedance of traces balanced and short to minimize reflections and functionality degradation (4.8).
14. Design clock traces as transmission lines to minimize or prevent reflections, ringing, and creation of RF common-mode currents (4.8).
15. Measure the actual routed trace length of all clocks and periodic or high-threat signals. Use Eqs. (4.14) through (4.17). Determine if the actual routed length is longer than maximum permissible calculated length. If so, termination is required (4.9).
16. If traces must be electrically long, route the trace using transmission line theory (4.9).
17. Terminate all clock traces in their characteristic impedance (4.9).
18. Route clock traces on one routing plane only. This layer must be adjacent to a solid (image) plane at all times. If possible, route all clock traces stripline. Traces on the bottom of the board are still microstrip (4.10).
19. Do not layer jump clock or high-threat traces between different layers. Doing so disrupts RF coupling between the trace and an image plane. Disruption prevents RF return current from completing its route uninterrupted from source to load in a continuous manner. If traces must jump between planes, use ground vias at each and every layer jump to maintain image plane continuity (4.10).
20. Microstrip allows for fastest transition of signal edges while permitting greater amount of RF currents to be radiated from the trace (4.10).
21. Stripline allows for optimal suppression of RF currents, but at the expense of slowing down signal edges (in the picosecond range) due to capacitive loading between the trace and surrounding planes (4.10).
22. Place a guard trace around each and every clock trace if the board is single or double sided (no ground plane present). Make the distance spacing as close as possible, with minimal distance between the signal and guard trace, and otherwise obey the 3-W rule. This minimizes crosstalk and provides a return path for RF current (4.11 and 4.16).
23. For high-threat signals, use of a shunt trace provides additional RF suppression by providing an additional ground return path for

common-mode currents that may exist on a trace in a board with dual stripline topology (4.11).

24. When using guard and shunt traces, make connections to the ground planes at irregular intervals throughout the route. Symmetrical grounding allows for a tuned circuit to be created that may be resonant at a particular harmonic or wavelength of a particular clock signal (4.11).
25. Use of both guard and shunt traces allows for a coaxial based transmission line to exist in the middle of the board (4.11).
26. Do not route two different signals between the same guard traces—crosstalk could develop. If the traces are paired (differential), then only these two traces may be routed with the same guard trace (4.11).
27. Crosstalk may be eliminated or reduced by guard traces or routing 3-W (4.12).
28. Another technique to prevent crosstalk is to route parallel traces separated by 2 mils/inch (0.002"/inch) of trace spacing (4.12).
29. Route all clock traces in a radial manner. Do not daisy-chain. Provide a series resistor for each radial trace with a fan-out of one device per driver, if possible (4.13).
30. Calculate the series resistor to be greater than or equal to the source impedance of the driving component and lower than or equal to the line impedance (4.13).
31. Do not use stubs or "T" connections on clock signals unless electrically short (4.13).
32. Calculate the value of the decoupling capacitor for individual traces based on the impedance of the circuit/trace and self-resonant frequency of the network. Guarantee that the edge rate of the signal is not degraded to the point of nonfunctionality (4.14).
33. Use oscillators for frequencies above 5 MHz or clock skews faster than 5 ns instead of discrete components or crystals (4.1 and 4.15).
34. Make provision for additional grounding of oscillator cases (4.1 and 4.15).
35. Crosstalk may be eliminated by routing traces 3-W. This rule states that "the distance separation between these traces must be three times the width of a single trace, from centerline to centerline. For dual stripline, one trace is three times the width of its corresponding trace (4.16).

A.4 INTERCONNECTS AND I/O

1. Provide EMI and ESD protection on all interconnects and I/O. This includes front panel display indicators and controls, I/O interconnects, power cords, blank adapter brackets, peripheral drive cover plates, interface devices, and the like.
2. Locate drivers and control logic as close to the I/O connector as possible to minimize trace length and RF coupling of both common- and differential-mode currents. Place filtering components between control logic and the I/O connector.
3. Bond all metal I/O connector housings 360° to chassis ground.
4. Physically partition interconnects and I/O circuitry away from high-RF-bandwidth areas, especially CPU sections and fast control logic.
5. Make provisions for quiet areas by separating digital logic from analog circuitry along with their respective power and ground planes (5.1).
6. In addition to quiet areas, provide each and every I/O port with an isolated (quiet) ground and/or power planes (5.1).
7. Make provision in the artwork for inclusion of a fence to prevent internal radiated RF noise coupling between functional sections. Use the fence if necessary for EMI compliance or to enhance system performance. Decouple all fence ground connections to the ground planes using decoupling capacitors (5.1).
8. Isolate noisy and quiet areas through use of a partition or moat. A moat is an absence of copper on all layers—power, ground, and signal. Connection between the moated areas is accomplished using common-mode chokes (data line filters), isolation transformers, or a bridge. Route only those traces associated with this quiet area through the bridge or data line filters. Ground both ends of the bridge (if possible when using multipoint grounding) to chassis ground with a screw or equivalent means. Grounding removes high-frequency RF currents present on the power planes due to RF ground noise generated by voltage gradients between partitioned sections (5.2).
9. Use data line filters, ferrite devices, or isolation transformers for connection between noisy and quiet areas (5.2).
10. Do not allow unnecessary inductance to be developed in both the signal and signal return traces. The signal return trace may also be a ground plane. This includes use of inductors and ferrite beads.

Make the ground return trace, if used instead of a ground plane, three times (3×) the width of the power trace (5.2).

11. Never violate or cross a moat with any trace. Route all traces associated to the isolated area through the bridge, if provided (5.2).
12. Partition each I/O subsection into a unique functional area. Create a separate section for serial, parallel, Ethernet, SCSI, video, audio, etc. (5.2).
13. Implement quiet areas between control logic, I/O subsection, and I/O connectors. This quiet area includes all power and ground planes (5.2).
14. Do not place active or non-I/O components in a quiet area (5.2).
15. Use capacitive and/or inductive data line filtering (capacitive for differential-mode, inductors for common-mode) on each and every I/O line. Place these components as close to the I/O connector as possible. Determine if the interwinding capacitance of inductors and data line filters will cause signal functionality problems and degradation (5.3).
16. Determine proper placement of bypass capacitors—before or after the data line filter. If located between the filter and I/O connector, select a capacitor with a minimum 1500 V rating to prevent damage during an ESD event (5.3).
17. Make provisions in the artwork for inclusion of bypass capacitors in the I/O circuit. Use these capacitors only if required for compliance or functional concerns; prepare for the future (5.3).
18. Ground all I/O brackets to chassis ground unless single-point or isolated grounding is required. Also, connect this I/O bracket to the ground planes of the printed circuit board. Provide for multiple connections from the ground planes of the board to the bracket. This minimizes the aspect ratio between ground points for optimal loop control. If no external I/O connections exist on the adapter board, isolate the chassis mounting bracket from signal ground (5.3).
19. Be aware of single and multipoint ground locations. Use to maximum advantage (5.3).
20. For local area networks, filter the data signals with common-mode chokes. Provide for complete isolation from the main printed circuit board using a moat. Determine if interwinding capacitance of the common-mode choke exceeds network specifications (5.4).
21. For video, provide a pi (π) filter between the video controller and the I/O connector. Locate this filter as close to the connector as

possible. Provide the isolated digital ground reference to analog ground through an inductor or ferrite device. Locate all analog traces and components exclusively over the analog isolated planes (5.5).

22. Partition audio interfaces into three areas: digital, analog, and audio. Make the digital-to-analog connection through a bridge located directly under the audio controller or as close to it as possible. Route all traces between digital and analog through this bridge, including analog power. Isolate the analog section from the audio section by a second moat and additional data line filters. Do not connect *audio* ground to *chassis* or *analog* ground. Do not connect signal return on unshielded audio cables to analog or chassis ground (5.6).
23. Provide fusing on all traces carrying ac or dc voltage that travels external to the unit. This is a requirement for product safety. Use a cartridge, pico, or PTC fuse (5.7).
24. For circuits that carry high levels of ac or dc voltage (> 42.2 V), maximize creepage and clearance distances to prevent an electrical hazard due to an abnormal fault condition that may occur. This is a requirement for product safety (5.8).

A.5 ELECTROSTATIC DISCHARGE PROTECTION

1. Provide ESD protection on all I/O signals (located directly at the I/O connector) using spark gaps, Tranzorbs™, high-voltage capacitors and/or RC/LC filters. Inductive components work better for ESD suppression than capacitive filters. Use of multilayer boards increase the immunity level of the assembly for ESD events.
2. Minimize ground loops using the following techniques:
   - Keep all power and ground traces as close as possible.
   - Keep signal lines as close as possible to ground circuits.
   - Use bypass capacitors throughout the board for both high and low self-resonant ESD frequencies.
   - Keep trace lengths short.
   - Fill in unused areas of the board with as much ground plane as possible. Via-connect these ground fill areas to chassis ground in as many places as possible.
   - Moat or partition ESD-sensitive components away from other

sensitive circuits.

- Make all chassis ground connections low impedance with tight bonding or securement means.
- Internal ground planes should surround every plated through-hole to minimize ground loops.

3. Install an ESD guard band (both top and bottom layers) around the periphery of the board to prevent ESD coupling into logic areas. Discharges that do not enter through I/O interconnects can still cause lockup. Ground the guard band every 1/2 inch along the perimeter to chassis ground. This provides a low-impedance path for ESD energy dissipation. Do not use solder mask on the guard band.
4. Separate noninsulated chassis ground from traces by a minimum of 0.22 cm (0.09").
5. Chassis ground traces (if used) must have a length-to-width ratio of 4:1 or less—the same as for any "bond strap."
6. Locate all logic and filter components as close as possible to the I/O connector.
7. If power and ground are implemented as a grid on the board, connect the traces together in as many locations as possible to minimize loop areas.
8. Provide input power to the board away from edges or areas least susceptible to ESD. Locate the power connector in the center of the board, if possible. Backplane power connectors are usually exempt from this requirement, as backplanes are generally located some distance from I/O connectors.
9. Use an image plane adjacent to each signal routing layer in ESD-sensitive areas.
10. Use a ground trace adjacent to each signal layer in ESD-sensitive areas.
11. Keep noninsulated circuits and components at least 2 cm (0.8") away from user-accessible areas, switches, or non-grounded metallic objects that the operator can touch.

A.6 BACKPLANES AND DAUGHTER CARDS

1. Select proper pin assignment to maintain ground loop control. A large number of ground pins minimizes crosstalk, reduces emissions, and enhances signal quality (7.1).

2. Maintain constant impedance for all traces throughout the backplane from source to load. Provide proper termination to optimize signal quality (7.1).
3. Use impedance-controlled connectors wherever possible (7.1).
4. Use a backplane with as many ground planes and pins as feasible. Always reference the backplane to the card cage either by direct ground-chassis connections or by the use of a chassis plane located internal to the backplane. Decouple power supply return planes and ac chassis plane to system (chassis) ground with bypass capacitors (7.2).
5. Design backplanes with alternating signal and ground traces. Do not bunch return grounds to a single set of multiple pins at opposite ends of the connector—large RF loop currents will be created, in addition to possible crosstalk between adjacent traces (7.2).
6. Backplanes generate differential-mode noise due to ground loops and insufficient grounding between traces. Place a ground trace adjacent to and around all clock traces (7.2).
7. Calculate and maintain proper impedance for all routing layers in the backplane (7.2).
8. Locate an image plane adjacent to each and every signal plane. Bond the image planes together in as many locations as possible with vias (7.3).
9. Determine if the top layer of the backplane is to be a ground plane or signal routing plane. Making this top layer a ground plane minimizes the impedance mismatch present between the backplane and I/O connectors and adapter cards (7.3).
10. If using many connector slots, perform a worst-case analysis to detect waveform degradation due to both lumped and distributed capacitance. Always provide for sufficient grounding of the backplane in multiple locations. Include in the measurement the adapter cards that plug into the backplane (7.3).
11. Verify that there are no high-threat or clock traces on the bottom routing layer of an adapter card that may couple to components and traces on the top layer of an adjacent adapter card (7.3).
12. Select interface connectors appropriate for the edge rate of the desired signal, along with proper impedance matching between the backplane and adapter cards (7.4).
13. When designing interconnects keep in mind the following (7.4):
    - Keep all discontinuities as short as possible.

- Use as many ground connections as possible within the allocated space or pinout.
- Establish a common ground within the connector.
- Use low-dielectric-constant board materials.
- Maintain the ground path length as close as possible to the path between both signal and ground.

14. Extend the physical dimension of the backplane (if possible) approximately one inch beyond the edge of the mounting bracket. Locate bypass capacitors and cable interconnects in this area. Ground the ac *chassis* plane to *chassis* ground in as many locations as possible at close intervals. Use bypass capacitors at each ground location (7.5).
15. Route all traces orthogonal between layers (horizontal versus vertical layer) (7.5).
16. Avoid vias between planes for all traces, clocks, and signals. All traces must be on the same plane, if possible. Do not daisy-chain clock or high-speed traces. Use radial clock distribution (7.6).
17. Do not use T-stubs in a backplane (7.6).
18. Make all traces as short as possible to prevent ringing and reflections that may occur on electrically long traces (7.7).
19. Terminate all signal and clock traces in their characteristic impedance. Make the terminating resistor the last item on the bus if end termination is used. Use the slowest logic possible. Design all traces as transmission lines (7.7).
20. To minimize crosstalk between traces and planes, use the 3-W rule or separate parallel traces at 0.002"/inch (2 mils) per inch of trace length (7.8).
21. For one- or two-layer backplanes, route ground traces in parallel to each signal trace to minimize ground loops and crosstalk that may occur between traces that are electrically long (7.9).
22. Use a signal return ground trace between parallel traces in a backplane. Tie this trace to the main system ground (7.9).
23. Do not route signal traces across through-hole connector pins that overlap. This routing method prevents the image planes from providing a low-impedance path for RF return currents. Return currents must travel around the long edge of the I/O connector to complete their path, thus maximizing generation of RF common-mode noise (7.10).

A.7 ADDITIONAL DESIGN TECHNIQUES

Trace Routing for Corners

1. Do not use 90° angles when routing traces with fast edge time components (8.1).

Selection of a Ferrite Device

1. When selecting a ferrite device for suppression of RF energy, take into consideration the following (8.2):
   - Select a ferrite material based upon the impedance presented to the circuit.
   - Determine the permeability value of the ferrite material for optimal frequency range of operation.
   - Alter the core size, shape, or length to change the impedance value of the ferrite device.
   - Elevated temperatures will decrease overall impedance and performance.
   - Extreme bias (current flow through the material) decreases the impedance to the point of nonfunctionality.
   - Determine whether ac or dc current is passed through the device. Excessive attenuation of a "wanted" signal could occur.
   - To increase the impedance of shield beads or cores, additional turns of wire may be added.

Grounded Heatsinks

1. On high-speed VLSI processors (generally 75 MHz and above), a grounded heatsink may be required (8.3).
   - Connect the heatsink to the ground planes through a mounting fence or bracket.
   - Decouple the heatsink's ground posts with alternate sets of parallel decoupling capacitors around all four sides of the processor.
   - Select a dielectric material for thermal conduction of heat. This material is also used for determining the self-resonant frequency of the heatsink assembly as a common-mode decoupling capacitor.

Lithium Battery Circuits

1. Provide reverse bias protection for all applications that use lithium batteries, per product safety requirements. This includes discrete batteries, nonvolatile RAM, clock calendars, etc. (8.4).

BNC Connectors

1. Ground the shell of the BNC connector to chassis ground at the bulkhead or nearest ground located on the printed circuit board. Provide a low-impedance path to chassis ground for RF currents present on the shield of the coax (8.5).
2. If using an isolated BNC connector, do not bond the shell or shield of the connector to chassis ground. Isolate the base of the connector using a nonconductive bushing (insulator) or equivalent means. Be certain to provide bypass EMI capacitors and ESD protection (8.5).
3. Do not use pigtails under any condition to connect the shell or ground pin of the BNC connector to chassis ground or to any other ground in the system (unless system functionality mandates this ground connection) (8.5).

Film

1. Provide impedance-controlled test coupons to determine quality of manufacturing from the vendor and to guarantee compliance with fabrication specifications.
2. Use a layer stackup window to determine number or layers internal to the board and to quickly determine which layers are signal, power, and ground.
3. Install test points for ease of test and debug.
4. Ensure that the thickness of the copper layers is appropriate for proper power distribution and to minimize ground bounce.
5. Do not place solder mask on any ground connections. This includes ground points, I/O bulkhead connectors, I/O adapter bracket mounting holes, ESD guard band, and clock circuit ground pads and localized ground plane on both component (top) and circuit (bottom) layers.
6. Remove unconnected vias from artwork prior to film. Unconnected vias are interconnects on planes that are not used by any trace. If a guard trace is routed around a via, make sure that there is no connection to that pad on that layer.
7. Specify appropriate fabrication material using flame-retardant fiberglass or equivalent material per product safety standards—generally V-1 minimum.

Appendix B

International EMI Specification Limits

Note: These requirements are subject to change at the discretion of various regulatory agencies. The reader is urged to verify the specifically applicable and current requirements that are in force at the time of product design and release.

B.1 DEFINITION OF CLASSIFICATION LEVELS

The FCC (United States of America) and DOC (Canada) use the same definitions:

Class A:

A digital device that is marketed for use in a commercial, industrial, or business environment, exclusive of a device that is marketed for use by the general public or is intended to be used in the home.

Products are self-verified for compliance.

Class B:

A digital device that is marketed for use in a residential environment, notwithstanding its use in a commercial, industrial, or business environment.

Products require Certification from the Federal Communication Commission (FCC). Canada accepts FCC Certification.

International definition, defined in EN 55 022:

Class A:

Equipment is information technology equipment that satisfies the Class A interference limits but does not satisfy the Class B limits. In some countries, such equipment may be subjected to restrictions on its sale and/or use.

(Note: The limits for Class A equipment are derived for typical commercial establishments for which a 30 m protection distance is used. The class A limits may be too liberal for domestic establishments and some residential areas).

Class B:

Equipment is information technology equipment that satisfies the Class B interference limits. Such equipment should not be subjected to restrictions on its sale and is generally not subject to restrictions on its use.

(Note: The limits for Class B equipment are derived for typical domestic establishments for which a 10 m protection distance is used).

B.2 FCC/DOC EMISSION LIMITS

For FCC and DOC compliance, the frequency range is based on the highest fundamental internally generated clock frequency per the list below.

< 1.705 MHz	Test to 30 MHz
1.705 MHz to 108 MHz	Test to 1 GHz
108 MHz to 500 MHz	Test to 2 GHz
500 MHz to 1 GHz	Test to 5 GHz
>1 GHz	Test to 5th harmonic or to 40 GHz, whichever is lower

FCC/DOC Class A Radiated Emission Limits

Frequency (MHz)	*Distance (m)*	*Quasipeak Limit (dBμV/m)*
30 to 88	10	39.0
88 to 216	10	43.5
216 to 960	10	46.5
>960	10	49.5

FCC/DOC Class A Conducted Emission Limits

Frequency	*Quasipeak Limit*
0.45 to 1.705 MHz	60.0 dBμV
0.705 to 30.0 MHz	69.5 dBμV

FCC/DOC Class B Radiated Emission Limits

Frequency (MHz)	*Distance (m)*	*Quasipeak Limit (dBμV/m)*
30 to 88	3	40.0
88 to 216	3	43.5
216 to 960	3	46.0
>960	3	54.0

FCC/DOC Class B Conducted Emission Limit

Frequency	*Quasi-Peak Limit*
0.45 to 30.0 MHz	48.0 dBμV

Table B.1 Summary of FCC/DOC Limits

	FCC/DOC Limits	
Frequency (MHz)	*A Limit*	*B Limit*
0.45–1.705	60 dBμV*	48 dBμV*
1.705–30	70 dBμV*	48 dBμV*
30–88	39 dBμV @ 10 m	40 dBμV @ 3 m
88–216	43.5 dBμV @ 10 m	43.5 dBμV @ 3 m
216–960	46.5 dBμV @ 10 m	46 dBμV @ 3 m
>960	49.5 dBμV @ 10 m	54 dBμV @ 3 m

*Narrowband limit—broadband limit is 13 dB higher.

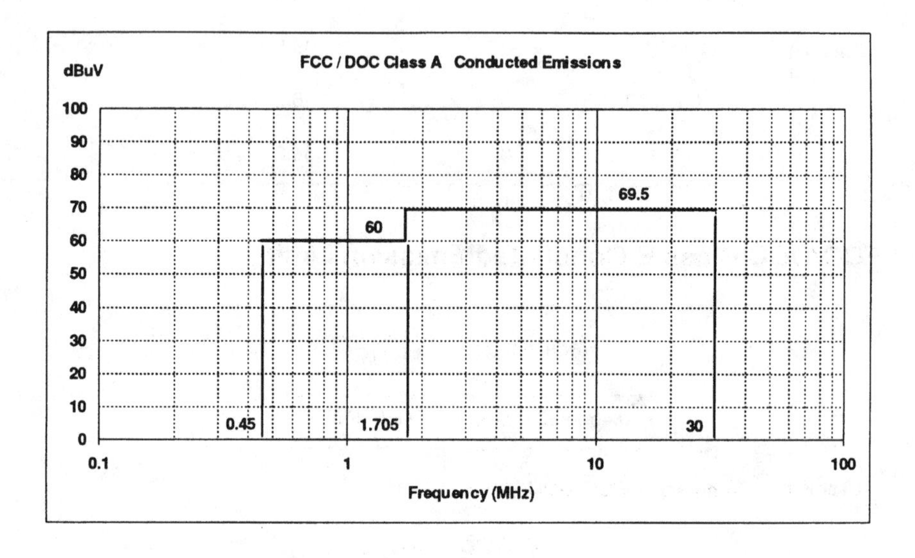
FCC / DOC Class A Conducted Emissions
dBuV
100
90
80
70
60
50
40
30
20
10
0
69.5
60
0.45
1.705
30
0.1
1
10
100
Frequency (MHz)

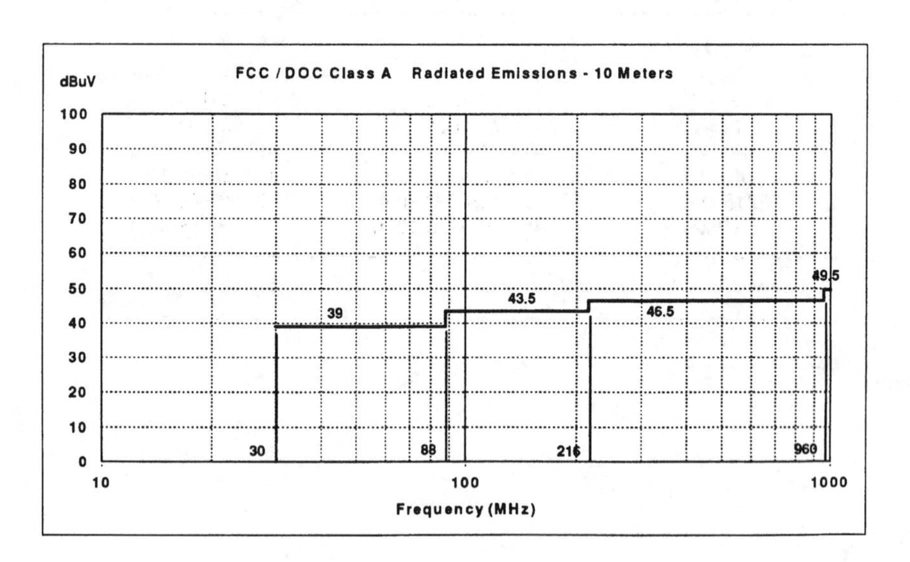
FCC / DOC Class A Radiated Emissions - 10 Meters
dBuV
100
90
80
70
60
50
40
30
20
10
0
49.5
43.5
39
46.5
30
88
216
960
10
100
1000
Frequency (MHz)

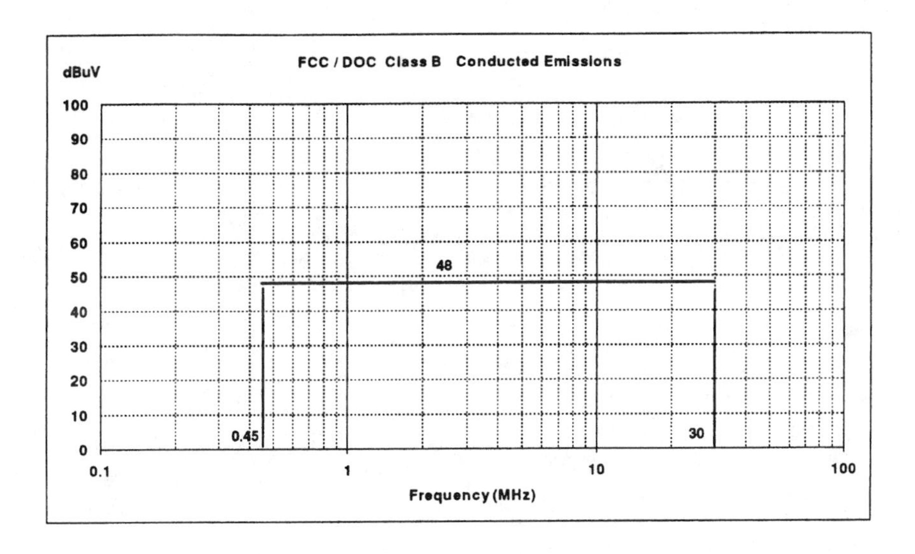
FCC / DOC Class B Conducted Emissions
dBuV
100
90
80
70
60
50
40
30
20
10
0
48
0.45
30
0.1
1
10
100
Frequency (MHz)

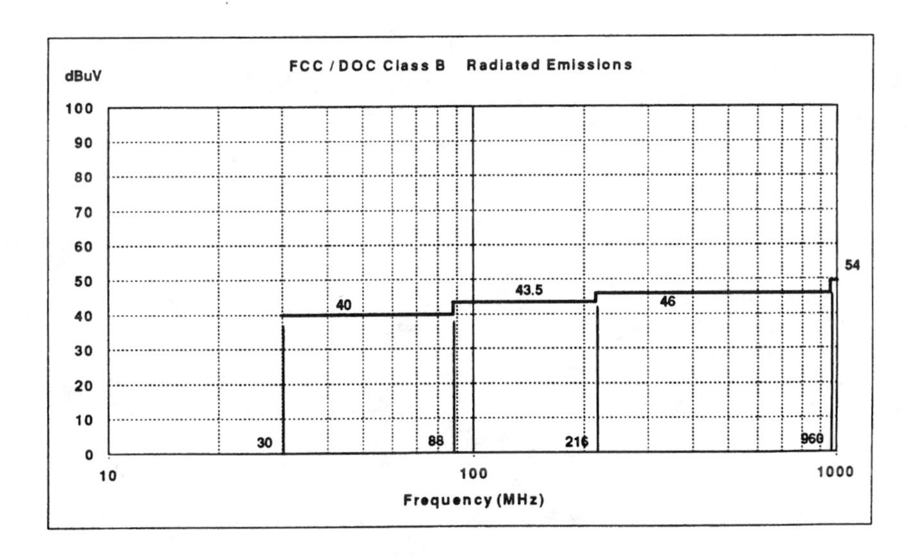
FCC / DOC Class B Radiated Emissions
dBuV
100
90
80
70
60
50
40
30
20
10
0
40
43.5
46
54
30
88
216
960
10
100
1000
Frequency (MHz)

B.3 INTERNATIONAL EMISSION LIMITS SUMMARY (SAMPLE LIST)

Table B.1 Summary of International Emission Limits

Class B Limits for Light Industrial Equipment and Primarily Residential Areas

	Frequency Range, MHz								
	0.15–0.5 dBμV		0.5–5 dBμV		5–30 dBμV		30–230 dBμV/m	230–1000 dBμV/m	
Specification	Quasipeak[1]	Average[1]	Quasipeak	Average	Quasipeak	Average	Quasipeak	Quasipeak	Notes
EN 50 081-1	66–56	56–46	56	46	60	50	30	37	@10 m, B limit
EN 55 011[2]	66–56	56–46	56	46	60	50	30	37	@10 m, B limit
EN 55 013[3]	66–56	56–46	56	46	60	50	45–55[4]	—	dBpW, absorbing clamp[4]
EN 55 014	66–56	56–46	56	46	60	50	45–55[4]	—	dBpW, absorbing clamp[4]
EN 55 020	66–56	56–46	56	46	60	50	45–55[4]	—	@ 10 m
EN 55 022	66–56	56–46	56	46	60	50	30	37	@ 10 m
Class A Limits for Industrial Areas									
EN 50 081-2	79	66	73	60	73	60	30	37	@ 30 m, A limit
EN 55 011[2]	79	66	73	60	73	60	30	37	@ 30 m, A limit
EN 55 022	79	66	73	60	73	60	30	37	@30 m
							40	47	@10 m

[1]The dash between two numbers (e.g., 66–56) means that the limit decreases with the logarithm of frequency.
[2]Detailed specification limits for EN 55 011 are shown in tables that follow.
[3]EN 55 013 has different limits for emissions from receivers and television sets.
[4]Absorbing clamp measurement is for the frequency range of 30–300 MHz only.

EN 55 011 is for equipment covered under EN 50 081-1 and is the product standard for Industrial, Scientific, and Medical Equipment. This standard defines limits for radiated and conducted emissions. EN 55 013 is for equipment covered under EN 50 081-1 for unintentional RF emissions from household electronics. EN 55 014 is for equipment covered under EN 50 081-1 for unintentional RF generators such as Burch motors and 50 Hz speed controls. This is also the emission product standard for Household Appliance Equipment (HHA). EN 55 020 is for equipment covered under EN 50 082-1 for immunity to RFI from household electronics. EN 55 022 is the emission requirement for products covered under EN 50 081-2 as well as Information Technology Equipment (ITE) and Electronic Data Processing (EDP) equipment.

EMISSIONS: EN 55 011 INDUSTRIAL, SCIENTIFIC, AND MEDICAL (ISM) EQUIPMENT

For all other EN 55 XXX specifications, refer to the previous table, Summary of International Emission Limits.

Classification of ISM Equipment

Group 1 ISM. Group 1 contains all ISM equipment from which there is intentionally generated and/or conductively coupled RF energy that is necessary for the functioning of the equipment itself.

Group 2 ISM. Group 2 embraces all ISM equipment in which RF energy is intentionally generated and/or used in the form of electromagnetic radiation for the treatment of material, and spark erosion equipment.

Line Conducted Emissions

These include emission levels lower than Class A limits (see Tables B.2 and B.3), or as agreed to with the competent body. The need for mains termination disturbance voltage limit for Class A equipment *in-situ* is under consideration.

Table B.2 Class A Equipment Limits dB(μV)

	Group 1		*Group 2**	
Freq. (MHz)	*Quasipeak*	*Average*	*Quasipeak*	*Average*
0.15–0.5	79	66	100	90
0.5–5	73	60	86	76
5–30	73	60	90 Decreasing with log of frequency to 70	80 Decreasing with log of frequency to 60

*Mains terminal disturbance voltage limits for Group 2, Class A equipment requiring currents greater than 100 A are under consideration.

Table B.3 Class B Equipment Limits dB(μV)

	Group 1	
Freq. (MHz)	*Quasipeak*	*Average*
0.15–0.5	66, decreasing with log of freq. to 56	56, decreasing with log of freq. to 46
0.5–5	56	46
5–30	60	50

Radiated Emissions

Table B.4 Emissions for Group 1 equipment

	Measured on a test site		*Measured in-situ*
Frequency band (MHz)	*Group 1, Class A 30 m (dBµV/m)*	*Group 1, Class B 10 m (dBµV/m)*	*Class A 30 m from exterior wall of building in which equipment is situated (dBµV/m)*
0.15–30	Under consideration	Under consideration	Under consideration
30–230	30	30	30
230–1000	37	37	37

Table B.5 Emissions limits for Group 2, Class B equipment measured on a test site

Frequency (MHz)	*Class B limits @ 10 m (dBµV/m)*
0.15–30	Under consideration
30–80.872	30
80.872–81.848	50
81.848–134.786	30
134.786–136.414	50
136.414–230	30
230–1000	37

Table B.6 Emissions limits for Group 2, Class A

	Limits with measuring distance 30 m	
Frequency (MHz)	*From exterior wall of building in which equipment is situated (dBµV/m)*	*Measured on a test site (dBµV/m)*
0.15–0.49	75	85
0.49–1.705	65	75
1.705–2.194	70	80
2.194–3.95	65	75
3.95–20	50	60
20–30	40	50
30–47	48	58
47–68	30	40
68–80.872	43	53
80.872–81.848	58	68
81.848–87	43	53
87–134.786	40	50
134.786–136.414	50	60
136.414–156	40	50
156–174	54	64
174–188.7	30	40
188.7–190.979	40	50
190.979–230	30	40
230–400	40	50
400–470	43	53
470–1000	40	50

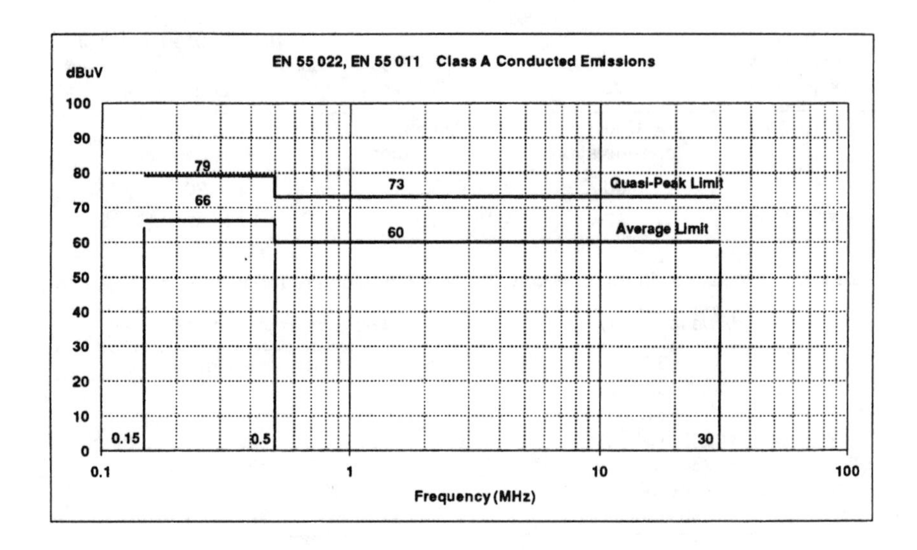
EN 55 022, EN 55 011 Class A Conducted Emissions
dBuV
100
90
80
70
60
50
40
30
20
10
0
79
66
73
60
Quasi-Peak Limit
Average Limit
0.15
0.5
30
0.1
1
10
100
Frequency (MHz)

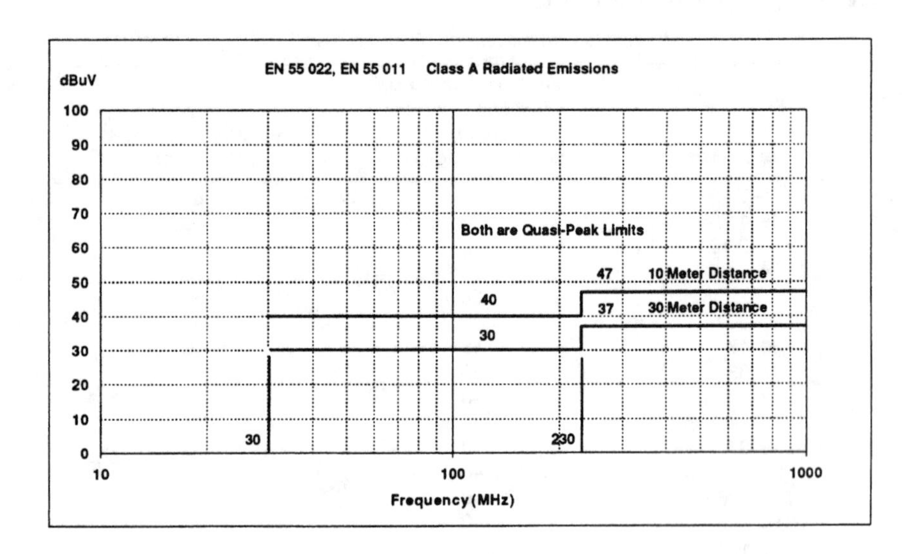
EN 55 022, EN 55 011 Class A Radiated Emissions
dBuV
100
90
80
70
60
50
40
30
20
10
0
Both are Quasi-Peak Limits
47
10 Meter Distance
40
37
30 Meter Distance
30
30
230
10
100
1000
Frequency (MHz)

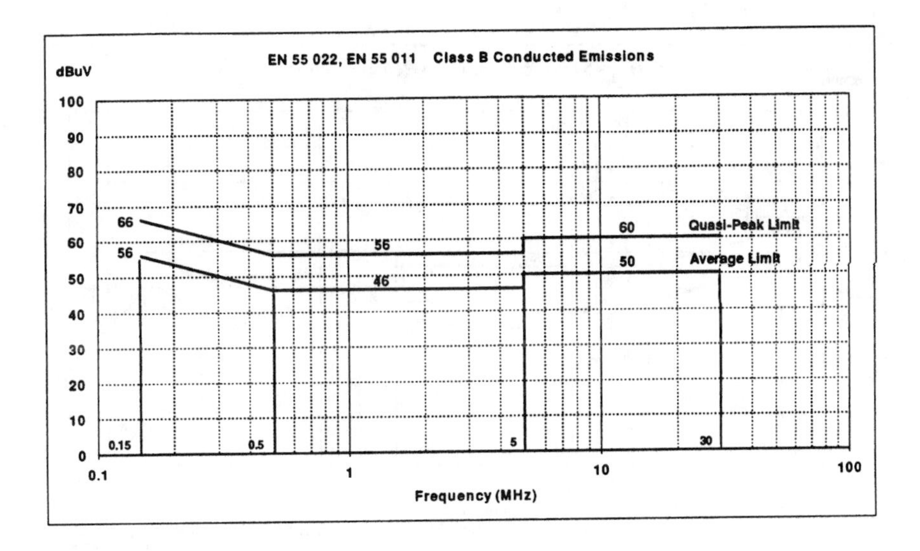
EN 55 022, EN 55 011 Class B Conducted Emissions
dBuV
100
90
80
70
60
50
40
30
20
10
0
66
56
56
46
60
50
Quasi-Peak Limit
Average Limit
0.15
0.5
5
30
0.1
1
10
100
Frequency (MHz)

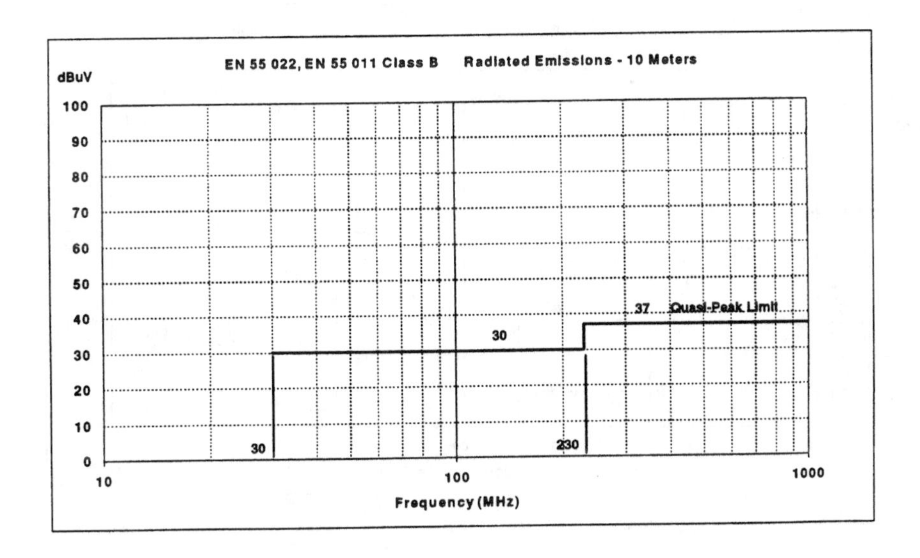
EN 55 022, EN 55 011 Class B Radiated Emissions - 10 Meters
dBuV
100
90
80
70
60
50
40
30
20
10
0
37
Quasi-Peak Limit
30
30
230
10
100
1000
Frequency (MHz)

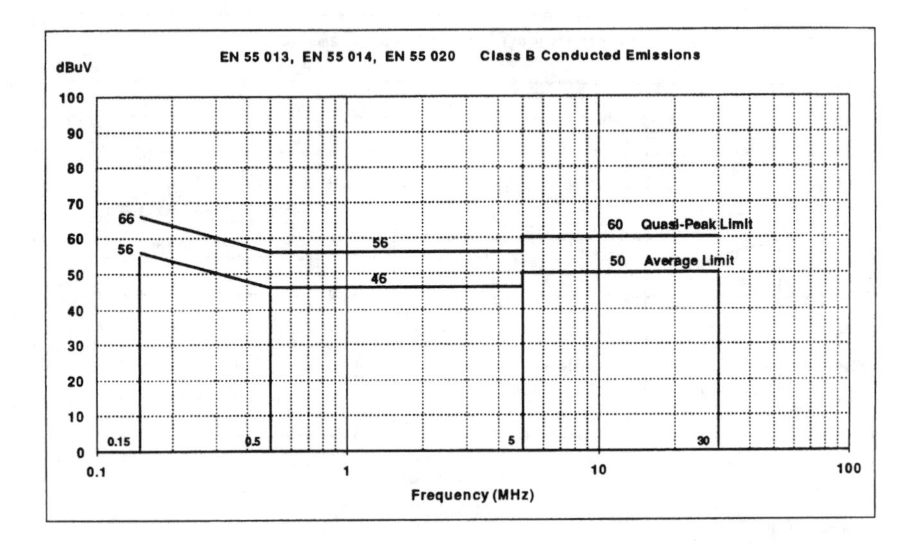
EN 55 013, EN 55 014, EN 55 020 Class B Conducted Emissions
dBuV
100
90
80
70
60
50
40
30
20
10
0
66
56
56
46
60 Quasi-Peak Limit
50 Average Limit
0.15
0.5
5
30
0.1
1
10
100
Frequency (MHz)

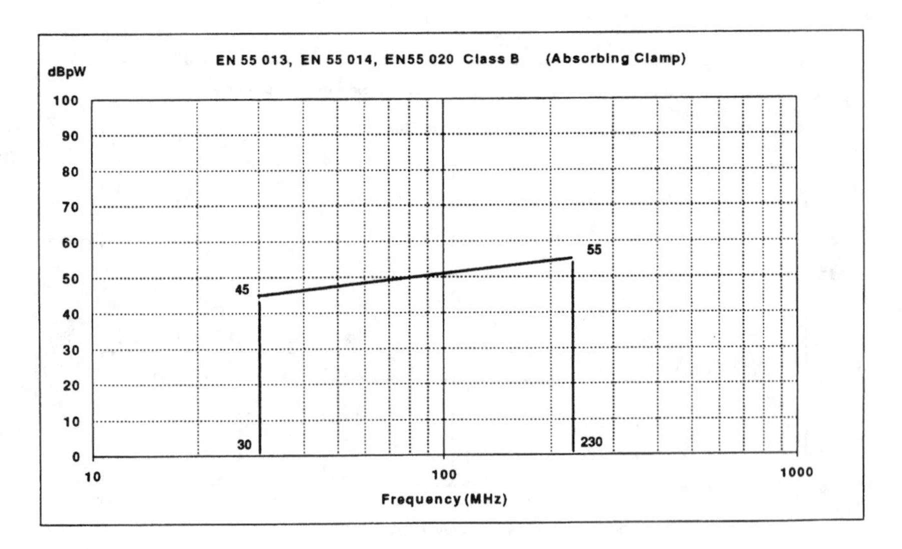
EN 55 013, EN 55 014, EN55 020 Class B (Absorbing Clamp)
dBpW
100
90
80
70
60
50
40
30
20
10
0
45
55
30
230
10
100
1000
Frequency (MHz)

Table B.7 International Immunity Requirements[1]

Specification	*IEC 1000-4-2 ESD*	*IEC 1000-4-3 Radiated RF Immunity*	*IEC 100-4-4 Electrical Fast Transients, AC Power*	*IEC 1000-4-5 Transients, Signal Leads*	*IEC 1000-4-6 Conducted RF Immunity*
EN 50 082-1, generic limit, light industrial equipment	8 kV (air) Criterion B	27–500 MHz 3 V/m Criterion A	500 V, signal 1000 V, AC 5/50 ns, 5 kHz Criterion B	Not yet proposed	Not yet proposed
EN 50 082-2, generic limit, heavy industrial equipment[2]	8 kV (air) 3 kV (direct) Criterion B	80–1000 MHz 10 V/m, plus 900 ± 5 MHz pulse modulate 200 Hz square wave Criterion A	1000 V, signal 2000 V, AC 5/50 ns, 5 kHz Criterion B	Not yet proposed	0.15–80 MHz 10 V 80% AM, 1 kHz 150 Ω source Criterion A
EN 55014-2, appliances and power tools	8 kV (air) 4 kV (direct) Criterion B	80–1000 MHz 3 V/m Criterion A	500 V, signal 1000 V, AC 5/50 ns, 5 kHz	1000 V, D.M. 2000 V, C.M. on power only 1.2/50 μs Criterion B	0.15–230 MHz Category II 0.15–80 MHz Category IV 1 V, signal 3 V, power Criterion A
EN 60601-2, medical devices	8 kV (air) 3 kV (direct) Criterion B	26–1000 MHz 3 V/m 80% AM, 1 kHz	500 V, signal 1000 V, AC 5/50 ns 5 kHz	1000 V, D.M. 2000 V, C.M. on power only 1.2/50 μs	Not yet proposed

[1]Severity levels and frequency ranges are subject to change. Consult test requirements for current values in effect at date of testing and certification.
[2]Additional test requirements exist that are not detailed above. Refer to EN 55 082-2 for details.

Performance criteria:
Level A: The apparatus shall continue to operate; no degradation of performance or loss of function is allowed.
Level B: The apparatus shall continue to operate as intended after the test.
Level C: Temporary loss of function is allowed, provided the loss of function is self-recoverable.

Bibliography

Boxleitner, W. 1988. *Electrostatic Discharge and Electronic Equipment.* New York: IEEE Press.

Brown, R., et al. 1973. *Lines, Waves and Antennas.* New York: Ronald Press Company.

Charles, J.P. 1985. Electromagnetic interference control in logic circuits. Proceedings of the 6th International EMC Symposium (Zurich, Switzerland), 145–150.

Condon, G.P. *Printed Circuit Board and Wiring Design for EMI Control.* West Conshohocken, Pa.: R&B Enterprises.

Coombs, C.F. 1988. Printed Circuits Handbook, 3d ed. New York: McGraw Hill.

Dash, G. 1989. Minimizing EMI at the PC board level. *Proceedings of EMC EXPO 89* (Washington, D.C.), T18.1–3.

Dockey, R.W. and R.F. German. 1993. Net techniques for reducing printed circuit board common-mode radiation. *Proceedings of the IEEE 1993 International Symposium on Electromagnetic Compatibility* (Dallas, Texas). New York: IEEE, 334–339.

Gabrielson, B., et al. 1989. Troubleshooting PC boards to reduce FCC emissions. *Proceedings of EMC EXPO 89* (Washington, D.C.), C1. 19–25.

Gavends, J.D. Measured effectiveness of a toroid choke in reducing common-mode current. 1989. *Proceedings of the IEEE 1989 National EMC Symposium* (Denver). New York: IEEE, 208–210.

Gerke, D.D., and W.D. Kimmel. 1987. Interference control in digital circuits. *Proceedings of EMC EXPO 87* (San Diego), T13.

Gerke, D.D., and W.D. Kimmel. 1994. *The designer's guide to electromagnetic compatibility.* EDN 39(2).

German, R.F., H. Ott, and C.R. Paul. Effect of an image plane on printed circuit board radiation. *Proceedings of the IEEE 1990 International Symposium on Electromagnetic Compatibility* (Washington, DC). New York: IEEE, 294–291.

German, R.F. 1985. Use of a ground grid to reduce printed circuit board radiation. *Proceedings of the 6th International EMC Symposium* (Zurich, Switzerland), 133–138.

Grasso, C. 1989. Printed circuit board design concepts for the control of EMC. *Proceedings of EMC EXPO 89* (Washington, D.C.). C1. 1–5.

Hartel, O. 1993. *Electromagnetic Compatibility by Design.* West Conshohocken, Pa.: R&B Enterprises.

Hejase, H.A.N., et al. 1989. Shielding effectiveness of pigtail connections. *IEEE Transactions on EMC* 31, 1:63–68.

Hnatek, E.R. No date. *Design of Solid-State Power Supplies,* 2d ed. New York: Van Nostrand Reinhold.

Hsu, T. 1991. The validity of using image plane theory to predict printed circuit board radiation. *Proceedings of the IEEE International Symposium on Electromagnetic Compatibility (*Cherry Hill, N.J.), 58–60.

Hubing, T. et al. 1995. Power bus decoupling on multilayer printed circuit boards. *IEEE Transactions on EMC* 37(2), 155–166.

Johnson, H.W., and M. Graham. 1993. *High Speed Digital Design.* Englewood Cliffs, NJ: Prentice Hall.

Keenan, R.K. 1983. *Digital Design for Interference Specifications.* Pinellas Park, Fla.: The Keenan Corporation.

Keenan, R.K. 1985. *Decoupling and Layout of Digital Printed Circuits.* Pinellas Park, Fla.: The Keenan Corporation.

Kendall, C. 1991. EMI Considerations for High Speed System Design. CKC Laboratories.

Khan, R.L., and G.I. Costache. 1984. Considerations of modeling crosstalk on printed circuit boards. *IEEE Transactions on MTT,* MTT-32, 7:705–710.

Kraus, John. 1984. *Electromagnetics.* New York: McGraw Hill.

Mardiguian, M. 1993. *Controlling Radiated Emissions by Design.* New York: Van Nostrand Reinhold.

McConnell, R. 1990. *Design Guideline for PCB Layout.* CKC Laboratories.

Motorola Semiconductor Products, Inc. 1990. Transmission Line Effects in PCB Applications, No. AN1051/D.

Montrose, M.I. 1991. Overview on design techniques for printed circuit board layout used in high technology products. *Proceedings of the 1991 International Symposium on Electromagnetic Compatibility* (Cherry Hill, NJ). New York: IEEE, 61–66.

Nave, M.J. 1986. Prediction of conducted emissions in switch power supplies. *Proceedings of the IEEE International EMC Symposium* (San Diego), 167–173.

Nave, M.J. 1986. Switched mode power supply noise: Common-mode emissions. *Proceedings of EMC EXPO 86* (Washington, DC) T-11.

Ott, H. *Noise Reduction Techniques in Electronic Systems,* 2nd ed. New York: John Wiley & Sons.

Parker, C. 1992. *Choosing a Ferrite for the Suppression of EMI.* Fair-Rite Products Corp.

Paul, C.R. 1989. A comparison of the contributions of common-mode and differential-mode currents in radiated emissions. *IEEE Transactions on EMC* 31, 2:189-193.

Paul, C.R. 1980. Effects of pigtails on crosstalk to braided-shield cables. *IEEE Transactions on EMC,* EMC-22, 3:161–172.

Paul, C.R. 1982. Effectiveness of multiple decoupling capacitors. *IEEE Transactions on EMC,* EMC-34(2), 130–133.

Paul, C.R. 1985. Printed Circuit Board EMC. *Proceedings of the 6th International Symposium on EMC* (Zurich, Switzerland), 107–114.

Paul, C.R. 1986. Modeling and prediction of ground shift on printed circuit boards. *Proceedings of Inst. Elec. Radio Eng. EMC Symposium* (York, England), 37–45.

Paul, C.R. 1992. *Introduction to Electromagnetic Compatibility.* New York: John Wiley & Sons.

Paul, C.R., and K.B. Hardin. 1988. Diagnosis and reduction of conducted noise emissions. *Proceedings of the IEEE 1988 International EMC Symposium* (Seattle, WA.), 19–23.

Smith, D. 1993. *High Frequency Measurements and Noise in Electronic Circuits.* New York: Van Nostrand Reinhold.

Swainson, D. 1988. Radiated emission and susceptibility prediction on ground plane printed circuit boards. *Proceedings of the Institute of Electrical Radio Engineers EMC Symposium* (York, England), 295–301.

Thomas, D.E., et al. 1983. Measurements and calculations of the crosstalk due to capacitive coupling between connector pins. *Proceedings of the IEEE 1983 International EMC Symposium* (Arlington, VA), 567–572.

Violette, J.L.N., et al. 1989. An Introduction to the design of printed circuit boards (PCBs) with high speed digital and high frequency system performance consideration. *Proceedings of EMC EXPO 89* (Washington, D.C.), A3.1–23.

Violette, J.L.N., et al. EMI control in the design and layout of printed circuit boards. *EMC Technology* 5, 2:19–32.

Violette, J.L.N., et al. 1987. *Electromagnetic Compatibility Handbook.* New York: Van Nostrand Reinhold.

Walter, R.L. 1987. Specifying and designing equipment interface circuitry which meets EMC/EMI requirements. *Proceedings of EMC EXPO 87* (San Diego), T-14.

Weston, D. 1991. *Electromagnetic Compatibility—Principles and Applications.* New York: Marcel Dekker, Inc.

White, D.R.J. 1989. *EMI Control Methods and Techniques—Electromagnetic Interference and Compatibility.* Gainesville, Va.: ICT.

White, D.R.J. 1982. *EMI Control in the Design of Printed Circuit Boards and Backplanes,* 2d ed. Gainesville, Va.: ICT.

Index